# SUPERIOR

# SUPERIOR

## The Return of Race Science

**Angela Saini**

BEACON PRESS
BOSTON

Beacon Press
Boston, Massachusetts
www.beacon.org

Beacon Press books
are published under the auspices of
the Unitarian Universalist Association of Congregations.



22 21 20 8 7 6 5 4 3 [HC]
23 22 21 20 8 7 6 5 4 3 [PBK]

This book is printed on acid-free paper that meets the uncoated paper ANSI/NISO specifications for permanence as revised in 1992.

Text design and composition by Kim Arney

*Library of Congress Cataloging-in-Publication Data*

Names: Saini, Angela, author.
Title: Superior : the return of race science / Angela Saini.
Description: Boston : Beacon Press, 2019. | Includes bibliographical references and index.
Identifiers: LCCN 2018060780 (print) | LCCN 2019015864 (ebook) | ISBN 9780807076941 (ebook) | ISBN 9780807076910 (hardback) | ISBN 9780807028421 (paperback)
Subjects: LCSH: Race—Research, | Eugenics. | BISAC: SOCIAL SCIENCE / Discrimination & Race Relations. | SCIENCE / Philosophy & Social Aspects. | SOCIAL SCIENCE / Anthropology / Cultural.
Classification: LCC HT1506 (ebook) | LCC HT1506 .S25 2019 (print) | DDC 305.8/00723—dc23
LC record available at https://lccn.loc.gov/2018060780

*For my parents,*
*the only ancestors I need to know.*

# CONTENTS

# PROLOGUE

*In the British Museum is where you can see 'em*
*The bones of African human beings*
—*English Breakfast* by Fun-Da-Mental

I'm surrounded by dead people, asking myself what I am.

*Where* I am is the British Museum. I've lived in London almost all my life, and through the decades I've seen every gallery many times over. It was the place my husband took me on our first date, and years later, it was the first museum to which I brought my baby son. What draws me back here is the scale, the sheer quantity of artifacts, each seemingly older and more valuable than the last. I feel overwhelmed by the grandeur of it. But as I've learned, if you look carefully, there are secrets.

When you arrive for the first time, it's almost impossible to notice them, the finer detail obscured by the visitors in a rush to tick off every major treasure. You get swept away, a fish in a shoal. The museum doesn't focus on one object, or even a few. The point is all of it. So many valuable things brought together like this have an obvious story to tell, one skillfully constructed to remind us of Britain's place in the world.

Medical doctor and collector Sir Hans Sloane bequeathed the founding collection that became the British Museum in 1753. It would come to document the entire span of human culture, in time and in space. The British Empire was growing, and in the museum you can see how the empire builders envisioned their position in history. Britain framed itself as the heir to the great civilizations of Egypt, Greece, the Middle East, and Rome. Just look at the enormous colonnade at the entrance, completed in 1852, mimicking ancient Greek architecture. The neoclassical style we associate with this corner of central London owes itself to the belief that the British saw themselves as the cultural and intellectual continuation of the great Greeks and Romans. The same brand of architecture on Capitol Hill in Washington, DC, tells us that America's nation builders saw themselves this way, too.

Britain, this small island nation, once had the might to take all these treasures, these eight million precious objects from every corner of the globe, and transport them here. The inhabitants of Rapa Nui built the giant bust of Hoa Hakananai'a to capture the spirit of one of their ancestors, and the Aztecs carved the exquisite turquoise double-headed serpent as an emblem of authority, but these masterpieces are in this museum now. No one thing is more important than the museum itself. It is a testament to the audacity of power and wealth.

The history of the world as seen through British eyes was a simple one: a straight line from nearby cultures in North Africa and the Middle East to southern and western Europe. Walk past the white marble sculptures removed from the Parthenon in Athens even as they crumbled. Walk past the statues of Greek and Roman gods, their bodies considered the ideal of human physical perfection, and you're witness to this narrative. In 1798, when Napoleon conquered Egypt and a French army engineer uncovered the Rosetta Stone, allowing historians to translate Egyptian hieroglyphs for the first time, this priceless object was claimed for France. It remains one of the most important historical objects in the world, a jewel of antiquity. A few years after it was found, though, the British army captured it and brought it here, where it has remained ever since. You'll see that one side of the stone is still inscribed with the words "Captured in Egypt by the British Army." As historian Holger Hoock writes, "The scale and quantity of the British Museum's collections owe much to the power and reach of the British military and imperial state." Know its history, and you begin to see the museum as a testament to the struggle for domination, to possess the deep roots of civilization itself.

Not long after Sloane bequeathed his collection, white European scientists also began to define what we now think of as race. In 1795, in the third edition of *On the Natural Varieties of Mankind*, a German doctor, Johann Friedrich Blumenbach, described five human varieties: Caucasians, Mongolians, Ethiopians, Americans, and Malays. To be precise, "Caucasian" refers to people who live in the mountainous Caucasus region between the Black Sea to the west and the Caspian Sea to the east, but under Blumenbach's sweeping definition it encompassed everyone from Europe to India and

North Africa. His arbitrary classification would have lasting consequences. We now use "Caucasian" as the polite way to describe white people.

But what does this mean today? Take the case of Mostafa Hefny, who considers himself very firmly and very obviously black. Authorities in the United States insist that he is white. He points to his skin, which is darker than that of some self-identified black Americans. He points to his hair, which is black and curlier than that of some black Americans. To any everyday observer, he's a black man. But according to the rules laid out by the US government in its 1997 Office of Management and Budget standards on race and ethnicity, people who originate in Europe, the Middle East, and North Africa are automatically classified as white. Since Hefny arrived from Egypt, he is officially white. In 1997, aged forty-six, Hefny filed a lawsuit against the United States government to change his official racial classification from white to black. His predicament still hasn't been resolved.

Now you might think of Hefny as being in a unique pickle, but in one way or another, most of us fall through a crack when it comes to defining race. What we are, this hard measure of identity, something so deep that it's woven into our skin and hair, a quality that nobody can really change, is actually harder to pin down than we think. My parents are from India, which means I am often described as Indian, Asian, or simply "brown." When I grew up in southeast London in the 1990s, those of us who weren't white would often be categorized politically as black. The National Union of Journalists still considers me a "Black member." But by Blumenbach's definition, being ancestrally north Indian makes me Caucasian.

Like Mustafa Hefny, then, I too am black, white, and other colors, depending on your definition. My race, which might seem so obvious to one person, may be quite another thing to the next. And this is because, centuries ago, people placed boundaries around populations and territory as casually as moving pieces on a chess board. The boundaries could have been placed anywhere, but now we squirm to fit into them or jostle our way out of them.

Ultimately what matters isn't necessarily where the lines are drawn, but what they mean. What does it *mean* to be black or white or something else, and why does it matter to us?

At the time these labels were devised, the meaning was clear. The power hierarchy had white people of European descent sitting at the top. They believed themselves to be the natural winners, the inevitable heirs of great ancient civilizations. There are still many today who look at the world and imagine that the imbalances and inequalities we see are natural, that white

Europeans have some innate superiority that allowed them to conquer and take the lead, and that they will have it forever. They imagine that only Europe could have been the birthplace of modern science, or that only the Europeans could have conquered the Americas. They imagine, as French president Nicolas Sarkozy said in 2007, that "the tragedy of Africa is that the African has not fully entered into history. . . . There is neither room for human endeavour nor the idea of progress." Or, as President Trump reportedly said in a White House meeting with lawmakers in 2018, that Haiti, El Salvador, and parts of Africa are "shithole countries."

The subtext is that history is over, the fittest have survived, and the victors have been decided. But of course, history is never over, and it is always more complicated than we think. In Sir Hans Sloane's time, accounts of the past were scribbled hastily, without the benefit of knowing about the remarkable Indus Valley civilization more than five thousand years old. We still know relatively little about this civilization, except that it had sophisticated cities and conducted trade using precise weights and measures. Sloane could have known little of the more recent Aztec and Inca empires in South America, which upon their discovery by Europeans destabilized the very meaning of civilization by proving that highly sophisticated societies emerged independently elsewhere. They came as such a shock that some to this day still believe their cities were the work of aliens.

In the British Museum ancient objects scream the truth silently. Take a walk up to the plaster cast of a relief from the temple of Beit el-Wali in Lower Nubia, built by the pharaoh Ramesses II, who died in 1213 BCE. It's high, near the ceiling, and spans almost the entire gallery. See the pharaoh depicted as an impressive figure on a chariot, wearing a tall blue headdress and brandishing a bow and arrow, his skin painted burnt ochre. He is plowing into a legion of Nubians, dressed in leopard skins, some warriors painted black and some painted the same ochre as him. He sends their limbs into a tangle before they're finally conquered. As the relief shows, the Egyptians at that time believed themselves to be a superior people with the most advanced culture, imposing order on chaos. The racial hierarchy, if that's what you want to call it, looked this way in this time and place.

Then things changed. Downstairs on the ground floor is a granite sphinx from a century or two later, a reminder of the time when the Kushites, an ancient Nubian kingdom located in present-day Sudan, invaded Egypt. There was a new winner now, and the ram sphinx protecting King Taharqa illustrates how this conquering force took Egyptian culture and appropriated it. The Kushites built their own pyramids, in the same way that the British replicated ancient Greek architecture. Taharqa was a black king of Egypt. Through objects like this, one can see how power balances shifted throughout history. They reveal a less simple version of the past, of who we are. And it's a version that demands humility, warning us that knowledge is not just an account of what we know, but has to be understood as something shaped by those who happen to hold power when the account was written. A hundred years is nothing; everything can change completely within a millennium.

The Ancient Egypt galleries of the British Museum are always the most crowded, especially the small selfie-worthy space directly in front of the Rosetta Stone. What we don't think as we walk past the mummies in their glittering cases is that this is also a mausoleum. We're surrounded by the skeletons of real people who lived in a civilization no less remarkable than the ones that followed or that went before. Every society that happens to be dominant comes to think of itself as being the best, deep down. The more powerful we humans become, the more our power begins to be framed as natural as well as cultural. We paint our enemies as ugly foreigners and our subordinates as inferior. We invent hierarchies, give meaning to our own racial categories. One day, five thousand years forward, in another museum in another nation, these could be European or American bones encased in glass, what were once considered advanced societies replaced by new ones. History is never over. No place or people has a claim on superiority.

Race is the counter-argument. Race is at its heart the belief that we are born different, deep inside our bodies, perhaps even in character and intellect, as well as in outward appearance. It's the notion that groups of people have certain innate qualities that not only are visible at the surface of their skins but also run down into their innate capacities, that perhaps even help define the passage of progress, the success and failure of the nations our ancestors came from.

And it's so tempting to feel this. Many of those who come to the museum for the first time—I can tell you this from having spent hours watching them—are looking for their own place in these galleries. The Chinese tourists go straight to the Tang dynasty artifacts; the Greeks, to the Parthenon marbles. The first time I came here, I made a beeline for the Indian galleries. My parents were born in India, as were their parents, and theirs before them, so in the museum's Indian galleries was where I imagined I would find the objects most relevant to my personal history. All visitors have the same curious desire to know who their ancestors were, to know what *their* people achieved. We want to see ourselves in the past, forgetting that everything in the museum belongs to us all as human beings. We are each products of it all.

But, of course, that's not the lesson we take away, because that's not what the museum was designed to tell us. Objects here are trapped inside glass cabinets, under tight security as though any of them might dare to leap back thousands of miles to where they were created. Why are they in these rooms, and not where they were first shaped, built, painted, carved, or erected? Why do they live inside this museum in London, its neoclassical columns today stretching into the wet and gray sky? Why are the bones of Africans here, and not where they were buried, in the magnificent tombs that were created for them, where they were supposed to live out eternity?

Because this is how power works. It takes, it claims, and it keeps. It makes you believe that this is where the objects belong. It's designed to put you in your place.

I was once told of an elderly man in Bangalore, in south India, who ate his chapatis with a knife and fork because this was how the British ate. These notions of superiority and inferiority impact us all. When my great-grandfather fought in the First World War for the British Empire and when my grandfather fought in the Second World War for the British Empire, their contributions were forgotten, like those of countless other Indian soldiers. They were never considered strictly equal to their white counterparts. This is how it was. When boys from my school threw rocks at me and my sister when I was ten years old, telling us to go home, this is how it was.

The global power balance, as it played out in the eighteenth century, meant that treasures from all over the world could and would only end up in a museum like this one, because Britain was one of the strongest nations at the time. It, along with other European powers, were the latest colonizers, the most recent winners. So they gave themselves the right to take things. They gave themselves the right to document history their way, to define the scientific facts about humankind. Just as the United States would later, when it became the global superpower. Throughout, white thinkers told us that their cultures were better, that they were the proprietors of thought and reason, and they married this with the notion that they belonged to superior races. These became our realities.

The truth is something else.

CHAPTER 1

# Deep Time

*Are we one human species, or aren't we?*

I am on a road dotted with the corpses of unlucky kangaroos, just under two hundred miles inland from the western Australian city of Perth—at the other end of the world from where I call home. It feels like a wilderness. Everything is alien to my eyes. Birds I've never seen before make sounds I've never heard before. The dead branches of silvery trees, skeleton fingers, extend out of crumbly red soil. Gigantic rocks weathered over billions of years into soft pastel blobs resemble mossy spaceships. I imagine I've been transported to a galaxy beyond time, one in which humans have no place.

Except that inside a dark shelter beneath one undulating boulder are handprints.

Mulka's Cave is one of many ancient rock-art sites dotted across Australia, but this one is unique in this particular region for being so densely packed with images. I have to crouch to enter, navigating the darkness. One hand is all I see at first, stenciled within a spray of red ochre illuminated on the granite by a diffuse shaft of light. My eyes adjust, more hands appear. Infant hands and adult hands, hands on top of hands, hands all over the ceiling—hundreds of them in reds, yellows, oranges, and whites. As they become clearer in the half-light, it's as though they're pushing through the walls for a high five. There are parallel lines, too, maybe delineating the vague outline of a dingo.

The images are hard to date. Some may well be thousands of years old; others, very recent. What is known is that the creation of rock art on this continent goes back to what in cultural terms feels like the dawn of time. Following excavation at the Madjedbebe rock shelter, in Arnhem Land in northern Australia in 2017, the duration of modern human presence here was set conservatively at around 60,000 years—far longer than members of our species have lived in Europe, and long enough for people here to have witnessed an ice age, as well as the extinction of the giant mammals. And they may have been making art at the outset. At the Madjedbebe site, I'm told by one archaeologist who worked there, researchers found ochre "crayons" worked down to a nub. At another Australian site, this one 42,000 years old, there is evidence of ceremonial burial, bodies sprinkled with ochre pigment that would have to have been transported there over hundreds of miles.

"Something like a handprint is likely to have many different meanings in different societies and even within a society," I'm told by Benjamin Smith, a British-born rock-art expert based at the University of Western Australia. It may signify place, possibly to assert that someone was here. But determining meaning is not always simple. The more experts like him have tried to decipher ancient art, wherever it is in the world, the more they've found themselves only scratching the surface of systems of thought so deep that Western philosophical traditions can't contain them. In Australia a rock isn't just a rock. The relationship that indigenous communities have to the land, even to inanimate natural objects, is practically boundless—everyone and everything is intertwined.

What at first looks to me to be an alien wilderness isn't wild at all. It's a home that is more lived in than any other that I can imagine. Countless generations have absorbed and built upon knowledge of food sources and navigation. They have shaped the landscape sustainably over millennia, built a spiritual relationship with it and its unique flora and fauna. As I learn slowly, in the thinking of Aboriginal Australians, individuals seem to melt away in the world around them. Time, space, and object take on different dimensions. And none except those who have grown up immersed in this culture and place can quite understand. I know that I could spend the rest of my life trying to fathom it and get no further than I am now, standing lonely in this cave.

We can't inhabit minds that aren't our own.

I was a teenager before I discovered that my mother might not actually know her own birthdate. We were celebrating her birthday on the same day we always did in October when she told us in passing that her sisters thought she had actually been born in the summer. Pinning down dates wasn't routine when she was growing up in India. It surprised me that she didn't care, and my surprise made her laugh. What mattered to her instead was her intricate web of family relationships, her place in society, her fate as mapped in the stars. So I learned that the things we value are what we know. I'm obsessed by dates, but this is because I went to school in Britain. I compare every city I visit to London, where I was born. It's the center of my universe.

For archaeologists interpreting the past, deciphering cultures that aren't their own is the challenge. "Archaeologists have struggled for a long time to determine what it is, what is that unique trait, what makes us special," says Smith, who as well as working in Australia has spent sixteen years at sites in South Africa. It's work that has taken him to the cradles of humankind, where he has rummaged through the remains of the beginning of our species. And it's a difficult business. It's surprisingly tough to date exactly when *Homo sapiens* emerged. Fossils of people who shared our facial features have been found that date from 300,000 to 100,000 years ago. Evidence of art or at least the use of ochre is reliably available in Africa far further back than 100,000 years, before some of our ancestors began venturing out of the continent and slowly populating other parts of the world, including Australia. "It's one of the things that sets us apart as a species, the ability to make complex art," Smith says.

But even if our ancestors were making art a hundred millennia ago, the world then was nothing like the world now. More than forty thousand years ago there weren't just modern humans, *Homo sapiens*, roaming the planet, but also archaic humans, including Neanderthals (sometimes called cavemen because their bones have been found in caves) who lived in Europe and parts of western and central Asia. And there were Denisovans, we now know, another archaic human whose remains have been found in limestone caves in Siberia; their territory possibly extended to Southeast Asia and Papua New Guinea. There were also at various times in the past many other kinds of human, most of which haven't yet been identified or named.

In the deep past we shared the planet, even living alongside each other at certain times, in particular places. For some academics this cosmopolitan moment in our ancient history lies at the heart of what it means to be modern. When we imagine these other kinds of human, it's often as knuckle-dragging primitives. We *Homo sapiens* must have had qualities that they didn't have, something that gave us an edge, the ability to survive and thrive as they went extinct. The word "Neanderthal" has long been a term of abuse. Dictionaries define it both as an extinct species of human that lived in ice age Europe and as an uncivilized, uncouth man of low intelligence. Neanderthals and even *Homo erectus* made stone tools like our own species,

*Homo sapiens*, Smith explains, but as far as convincing evidence goes, he believes none had the same capacity to think symbolically, to talk in past and future tenses, to produce art quite like our own. These are the things that made us modern.

What separated "us" from "them" goes to the core of who we are. But it's not just a question for the past. Today being human might seem so patently clear, so beyond need for clarification, that we forget that once it wasn't so. The boundaries are still plagued by uncertainty. Scientific debate around what makes a modern human a *modern* human is as contentious as it has ever been. There are even quiet doubts about just how much the "same" all *Homo sapiens* living today really are. One old scientific theory claimed that, since we know there were other types of human alive tens of thousands of years ago in various parts of the world, maybe different races are in fact the descendants of these separate archaic forms?

From our vantage point in the twenty-first century, this might sound absurd. The common, mainstream view is that we have shared origins, as described by the "out of Africa" hypothesis. Scientific data has confirmed in recent decades that *Homo sapiens* evolved from a population of people in Africa before some of these people began migrating to the rest of the world around 100,000 years ago and then began adapting in small ways to their own particular environmental conditions. Within Africa, too, there was adaptation and change, depending on where people lived. Overall, however, modern humans were then and remain now one species, *Homo sapiens*. We are special. It's nothing less than a scientific creed.

But this view isn't shared universally within academia, nor is it even the mainstream belief in certain countries, including China. There are still scientists who ask, with perfectly serious faces, whether different populations actually evolved separately into modern humans—maybe leading to what we think of as racial difference. There are those who think that rather than modern humans migrating out of Africa, populations on each continent actually emerged into modernity separately from ancestors who lived there as far back as millions of years ago. They tell us we only need to travel into deep time to find our answers.

As unconscionable as it may seem, some suspect that population groups—perhaps equating to "races"—may have evolved into modern human beings in different ways.

| | | | | | | | |

In one early account of indigenous Australians by a European, a seventeenth-century English pirate and explorer, William Dampier, called them "the miserablest people in the world."

Dampier and the British colonists who followed him to the continent dismissed their new neighbors as savages who had been trapped in cultural stasis since they had migrated or emerged here, however long ago that was. Kay Anderson and Colin Perrin, cultural scholars based at Western Sydney University, document how the initial reaction of Europeans in Australia was sheer puzzlement. "The non-cultivating Aborigine bewildered the early colonists," they write. The Aboriginals didn't build houses, they didn't have agriculture, they didn't rear livestock. The colonists couldn't figure out why these people, if they were equally human, hadn't "improved" themselves by adopting these things. Why weren't the Aboriginals more like them?

There was more to this than culture shock. Bewilderment—or rather, an unwillingness to try and understand the continent's original inhabitants—suited Europeans in the eighteenth century because it also served the belief that they were entering a territory they could justly claim for themselves. The landscape was thought to be no different from how it must have been in the beginning, because they couldn't recognize how it might have been changed by the people living there. And if the land hadn't been cultivated, then by Western legal measures it was *terra nullius*—it didn't belong to anyone.

By the same token, if its inhabitants belonged to the past, to a time before modernity, their days were also numbered. "Indigenous Australians were considered to be primitive, a fossilized stage in human evolution," I'm told by Billy Griffiths, a young Australian historian who has documented the story of archaeology in his country, challenging the narrative that once painted indigenous peoples as evolutionarily backward. At least one early explorer even refused to believe they had created the rock art he saw. They were viewed as "an earlier stage of Western history, a living representative of an ancient form, a stepping stone." From almost the first encounter, Aboriginal Australians were judged to have no history of their own, to have survived in isolation as a flashback to how all humans might have lived before some became civilized. In 1958 the distinguished Australian archaeologist John Mulvaney wrote that Victorians saw Australia as a "museum of primeval humanity." Even at the end of the twentieth century, writers and scholars routinely called them Stone Age people.

It's true that these are cultures that have long connections to their ancestors, a continuation of traditions that go back millennia. "The deep past is a living heritage," Griffiths tells me. For Aboriginal Australians, "It's something they feel in their bones. . . . There are amazing stories of dramatic events that are preserved in oral histories, oral traditions, such as the rising of the seas at the end of the last ice age, and hills becoming islands, the eruption of volcanoes in western Victoria, even meteorites in different times." But this doesn't mean that ways of life have never changed. European colonizers failed to see this, and it would take until the second half of the twentieth century for that view to be corrected.

"There was certainly little respect for the remarkable systems of understanding and land management that indigenous Australians had cultivated over millennia," explains Griffiths. For thousands of years the land has been embedded with stories and songs, cultivated with digging sticks and fire and by hand. "While people have lived in Australia, there's been enormous environmental change as well as social change, political change, cultural change." Their lives have never been static. In his 2014 book *Dark Emu, Black Seeds* writer Bruce Pascoe argues, as other scholars have done, that this engagement with the land was so sophisticated and successful, including the harvesting of crops and fish, that it amounted to farming and agriculture.

But whatever they saw, the colonizers didn't value. Even now, for those raised in and around cities, industrialization is what represents civilization. Respect for and pride in indigenous cultures has only started to build in the last few decades, but even then, there is resistance among some nonindigenous Australians—especially as it has become clear from archaeological evidence that Aboriginal Australians have been occupying this territory not for thousands of years but for many tens of thousands. "The mid-twentieth century revelation that people were here for that kind of depth of time . . . was received in many ways as a challenge to a settler nation with a very shallow history. There are cultural anxieties wrapped up in all of this," says Griffiths. "It challenges the legitimacy of white presence here."

For those with a deeper sense of the past, Benjamin Smith says "the idea of ranking, say, an industrial society higher than a hunter-gatherer society is absurd." It's not easy to accept when you've grown up in a society that tells you concrete skyscrapers are the symbols of advanced culture, but when viewed from the perspective of deep time—across millennia rather than centuries, in the context of long historical trajectories—it becomes clearer. Empires and cities decline and fall. It is smaller, indigenous communities who survive throughout, whose societies date to many thousands rather than many hundreds of years old. "Archaeology shows us that all societies are incredibly sophisticated, they are just sophisticated in different ways," Smith continues. "These are the world's thinkers, and maybe they thought themselves into a better place. They have societies that have more leisure time than Western societies, lower suicide rates, higher standards of living in many ways, even though they don't have all of the technological sophistication."

Clearly, this wasn't the view of nineteenth-century European colonists. There was a failure to engage with those they encountered, to accept them as the true inhabitants of the land, combined with a mercenary hastiness to write them off. Like the native people of Tierra del Fuego at the southernmost tip of South America, whose nakedness and apparent savagery shocked biologist Charles Darwin when he saw them on his travels, indigenous Australians and Tasmanians were seen as occupying the lowest rungs in the human racial hierarchy. One observer described them as "descending to the grave." They were seen as doomed to go extinct, Griffiths tells me: "That was the dominant concept, that they would soon die out. There was a lot of talk of smoothing the pillow of a dying race."

"Smoothing the pillow" was bloodthirsty work. Disease was the greatest killer, the forerunner of invasion, he notes. But starting in September 1794, six years after the first fleet of British ships arrived in what was to become Sydney, and continuing into the twentieth century, hundreds of massacres also helped to slowly and steadily shrink the indigenous population by around 80 percent, according to some estimates. Many hundreds of thousands of people died, if not of smallpox and other illnesses brought to Australia, then directly at the hands of individuals or gangs and at other times of police. Equally harsh was the cultural genocide, says Griffiths. There were bans on the practice of culture and use of language. "Many people hid their identity, which also contributed to the decline in population."

In 1869 the Australian government passed legislation allowing children to be forcibly taken away from their parents, particularly if they had mixed heritage—described at the time as "half-caste," "quarter-caste," and smaller fractions. An official inquiry into the effects of this policy on the indelibly scarred "Stolen Generations," finally published in 1997, is a catalogue of horrors. In Queensland and Western Australia, for example, people were forced onto government settlements and missions, and children were removed from about the age of four and placed in dormitories, before being sent off to work at the age of fourteen. "Indigenous girls who became pregnant were sent back to the mission or dormitory to have their child," says Griffiths. "The removal process then repeated itself."

By the 1930s around half of Queensland's Aboriginal Australian population was living in institutions. Life was bleak, with high rates of illness and malnutrition, and the people's behavior strictly policed for fear that they would return to the "immoral" ways of their home communities. Children were able to leave dormitories and missions only to provide cheap labor, the girls as domestic servants and the boys as farm laborers. They were considered mentally unsuited to any other kind of work. The historian Meg Parsons describes what happened as the "remaking of Aboriginal bodies into suitable subjects and workers for White Queensland."

Among those forced to live this way were the mother and grandmother of Gail Beck, an indigenous activist in Perth who was once a nurse but now works at the South West Aboriginal Land and Sea Council, fighting to reclaim land rights for her local community, the Noongar. When I visit her at her home in the picturesque port city of Fremantle, speaking to her as she cooks while awaiting a visit from the Aboriginal Australian side of her family, I find someone who has few ways to quantify the pain and loss.

Gail is sixty years old, but her true family story is still fairly new to her. Until she was in her thirties, she didn't even know she had any indigenous ancestry. She had been raised to believe she was Italian—a lie to explain her olive skin, her mother terrified that if she told the truth, her daughter might be taken away by the authorities as she herself had been. So Gail lived under a conspiracy of silence, shielded from the fact that her grandmother had been one of the Stolen Generations, a "half caste" taken from her family to live in a Catholic missionary home in 1911 at the age of two. There, she had been abused, physically, mentally, sexually. "She was put out to service at thirteen. Didn't get paid, nothing like that. And she stayed there until she was an adult." A similar fate fell on Gail's mother, who was under the supervisory care of the nuns in the home from the day she was born, beaten and burned by them when she grew older. The Sisters of Mercy "were very cruel people," Gail recounts.

Learning about her family's past, and having it confirmed by her grandmother's papers, was a bolt from the blue. "I cried an ocean of tears." At once Gail gained a new identity, one that she was desperate to understand and build a connection to. It took her six years to find the part of her family that had been hidden from her, and she has devoted herself to absorbing their culture ever since. She shows me her blankets and pictures, adorned with the prints for which Aboriginal Australian artists have lately become famous. She has tried to learn an indigenous language, but it has been a struggle. She lives like most white Australians, in a nice house in a nice suburb, her knowledge of her great-grandmother's way of life, as it would have been, piecemeal.

"We are constantly in mourning, and people don't understand that," she tells me. "The young children that were lost, that doesn't just affect the nuclear family, that affects the community." And this is perhaps the greatest tragedy of all, that the way of life she might have had, the knowledge and language she could have been raised with, the relationship to nature, all of this was trodden beneath the boot of what considered itself to be a superior race. After the arrival of the Europeans, even the creation of art sharply declined. It took until 1976 for Aboriginal people even to be able to gain legal rights over their land. Throughout, the victims had no choice. "They weren't allowed to practice their culture, they weren't allowed to mix, and they weren't allowed to speak their language." Having been told they were inferior, that theirs was a life to be ashamed of, they adopted different ways of living—ways they were told were better.

"It was a real shameful thing."

I I I I I I I I I I

I don't cry easily. But in the car afterwards, I cry for Gail Beck. There is no scale of justice large enough to account for what happened. Not just for the abuse and the trauma, the children torn from their parents, the killings, but also for the lives that women and men like her didn't have the chance to live.

In recent decades, as scholars have tried to piece together the past and make sense of what happened, as they share with Australians in the long process of assessing the damage and its impact, we see an overarching story about how to define human difference. It is about where people have drawn boundaries around other groups of people, about how far inside us and how far back in time the disparities are thought to stretch. These are the parameters of what we now call race.

I meet with Martin Porr, a German archaeologist at the University of Western Australia whose work focuses on human origins. He feels, as do many archaeologists nowadays, that his profession is weighed down by the baggage of colonialism. When the first European encounters with Australians occurred, when the rules were drawn for how to treat indigenous peoples, science and archaeology began to be woven in. And they have remained interwoven ever since.

For Porr, this tale begins with the Enlightenment, at the birth of Western science. The Enlightenment reinforced the idea of human unity, of an essential biological quality that elevated humans above all other creatures. We live with this concept to this day, seeing it as positive and inclusive, a fact to be celebrated. There was a caveat, however. As Porr cautions, this modern universal way of framing human origins was constructed at a time when the world was a very different place: when European thinkers set the standard for what they considered a modern human, many built it around their own experiences and what they happened to value culturally at that time. To be fair, this was their lens through which everything was refracted in the same way that I compare every city to London. Those who lived in other lands, including the indigenous people of the New World and Australia, were at that time often a mystery to Europeans.

A number of Enlightenment thinkers, including influential German philosophers Immanuel Kant and Georg Wilhelm Friedrich Hegel, defined humanity without really having much of an idea how most of humanity lived or what it looked like. "A universal understanding of human origins was actually created at the time by white men in Europe who only had indirect access to information about other people in the world through the lens of colonialism," explains Porr. So when they went out into the real world and encountered people who didn't look like them, who lived in ways they didn't choose to live, the first question they were forced to ask themselves was: Are they the same as *us*? The problem was that, because of the narrow parameters they established of what constituted a human being, setting themselves as the benchmark, other cultures were almost guaranteed not to fit. In universalizing humanity by seeing themselves as the paradigm, they had laid the foundations for dividing it.

"If you define humanity in some universal sense, then it's very restrictive. And in the eighteenth century, that was totally Eurocentric. And of course, when you define it in that sense, then of course, so to speak, other people do not meet these standards," Porr says.

"When you look at these giants of the eighteenth century, Kant and Hegel, they were terribly racist. They were unbelievably racist!" Kant stated in *Observations on the Feeling of the Beautiful and the Sublime* in 1764, "The Negroes of Africa have by nature no feeling that rises above the trifling." When he met a quick-witted carpenter, he quickly dismissed him with the observation that "this fellow was quite black from head to foot, a clear proof that what he said was stupid." While a few Enlightenment thinkers did resist the idea of a racial hierarchy, many, including the French philosopher Voltaire and the English thinker David Hume, saw no contradiction between the values of liberty and fraternity and their belief that nonwhites were innately inferior to whites.

And here lay the flaw at the heart of modern science, one that would persist for centuries—arguably to this day. It is a science of human origins, as the British anthropologist Tim Ingold observes, that "has written the essence of humanity in its own image, and that measures other people by how far they have come in living up to it."

By the nineteenth century, those who didn't live like Europeans were thought not yet to have fully realized their potential as human beings. Even now, Porr notes that when scientists discuss human origins, he still catches them describing *Homo sapiens* as "better" and "faster" than and "superior" to other human species—easily interpreted as economic terms. There's an implicit assumption that higher productivity and more mastery over nature, the presence of settlements and cities, are the marks of human progress, even of the evolution of mankind. The more superior we are to nature, the more superior we are as humans. It is a way of thinking that still forces a ranking of people from closer to nature to more distant, from less developed to more, from worse to better.

History shows us that it's only a small leap from believing in cultural superiority to believing in biological superiority, that a group's achievements result from their innate capacities.

What Europeans saw as cultural shortcomings in other populations in the early nineteenth century soon became conflated with how they looked. The cultural scholars Anderson and Perrin explain how, in the nineteenth century, race came to be *everything*. One writer at the time noted that the natives of Australia differ "from any other race of men in features, complexion, habits and language." The fact they had darker skin and different facial features became markers of their separateness, a sign of their permanent difference. Their perceived failure to cultivate the land, to domesticate animals, and live in houses was taken part and parcel with their appearance. This had wider implications. Race, rather than history, could then be framed as the explanation for not only *their* failure, but for the failures of all nonwhite races to live up to the European ideal that Europeans themselves had defined. An Aboriginal Australian—just by virtue of having darker skin—could now be lumped together with a West African, despite their being continents apart and possessing different cultures and histories.

Whiteness became the visible measure of human modernity—an ideal that went so far as to become enshrined in Australian law. The historian Billy Griffiths explains: "When Australia federated in 1901, when the states came together as a nation, one of the first pieces of legislation to pass through Parliament was the Immigration Restriction Act, which formed the basis of the White Australia policy (strikingly similar to US policies against non-Northern European immigration around the same time). It sought to fuse the new nation together with whiteness by excluding non-European migration and attempting to assimilate and, ultimately, to eliminate Aboriginal and Torres Strait Islander identity." What happened to Gail Beck's family was part of these attempts to remove the color from Australia, in her case to breed it out of her mother's line over generations. "There was this horrible language of 'breeding out' the color from full-bloods to half-castes to quarter-castes to octoroons," Griffiths tells me. The goal was to steadily replace one "race" with another.

At the same time as this state-sanctioned ethnic cleansing was taking place, a crisis was emerging within scientific circles. For more than a century, most European thinkers had united around the Enlightenment idea that humankind was one, that we all shared the same common capacities, the same spark of humanity that made it possible for even those of us condemned as "miserable" by Europeans to improve, with enough encouragement. Even if there was a racial hierarchy, even if there were lesser humans and greater ones, we were all still *human*. But as Europeans encountered more people in other parts of the world, as they began to see the variety that exists across our species, and failed to "improve" people the way they wanted to, some began to seriously doubt this cherished belief.

Scientists ventured to wonder whether we all really did belong to the same species.

The course of the nineteenth century saw some take an intellectual shift away from the Enlightenment view of a single humanity with shared origins. This wasn't always just because of racism. Scientists had been funneled into a certain way of thinking about the world partly because of where they happened to be based. In the early days of archaeology, Europe was the reference point for subsequent research elsewhere. Before anyone was sure about humanity's African origins, human fossils in Europe provided the first data. According to John Shea, a professor of anthropology at Stony Brook University in New York, this created an indexing problem: setting European archaeological finds as the reference point for future discoveries, thus inadvertently placing Europe at the center of the story. "If you have a series of observations, the first observations guide you more so than the latter ones. And our first observations about human evolution were based on an archaeological record in Europe."

The first movements out of Africa were eastward, not westward. This is why you see elephants in both Africa and Asia, but not in Europe. Europe isn't where humans originated—in fact, its climate being so inhospitable back then, it was one of the last places humans migrated to, long after they arrived in Australia. But since Europe was where the first archaeologists happened to live and work, this geographical outpost became the model for thinking about our species' past.

Some of the very oldest human sites in Europe bear evidence of fairly sophisticated cave art. So as a result of indexing, early archaeologists digging on their doorstep logically assumed that representational art must be a mark of human modernity, one of the features that make us special. But the first *Homo sapiens* arrived in Europe only around 45,000 years ago. When researchers later excavated far earlier sites in Africa, some as old as 200,000 years, they didn't always find the same evidence of representational art. "The archaeologists came up with a way to square this," says Shea. "They said, well, okay, you know these ancient Africans, Asians, they look morphologically modern but they aren't behaviorally modern. They're not quite right yet." They decided that although they *looked* like modern humans, for some reason they didn't *act* like them.

Rather than rethinking what it meant to be a modern human—perhaps taking out the requirement that *Homo sapiens* must have had symbolic thought immediately upon the emergence of our species—researchers made the rest of the world's history a puzzle to be solved. It's a misstep that continues to have repercussions today. If representational art is what sets our species apart from Neanderthals and others, then at what point did we actually become our species? Did we do it 45,000 years ago when we see representational art in Europe, or 100,000 years ago, when we now know people used ochre for drawing, rather than 200,000 years ago? And if Neanderthals or other archaic humans turn out to have had representational art, will we then have to call them modern too? "Behavioral modernity is a diagnosis," says Shea. All the archaeologists can think to do is "rummage around looking for other evidence that will confirm this diagnosis of modernity."

In the nineteenth century this uncertainty around what constituted a modern human being took a leap further. If people weren't cultivating the land or living in brick houses, some asked, could they be considered modern? And if they weren't modern, were they even the same species?

Australia in all its strangeness posed a particular challenge to European thinkers. Anderson and Perrin argue that the discovery of the continent helped shatter the Enlightenment belief in human unity. After all, here was a remote place, with its own animals not seen elsewhere, kangaroos and koalas, and with its own plants and flowers and an alien landscape. "Based on observations of the uniqueness of Australian flora and fauna" there were "suspicions that the entire continent might have been the product of a separate creation," they write. The humans of Australia were thought to be as strange as everything else there.

Indeed, after Neanderthal remains were first discovered in 1856, Martin Porr and his colleague Jacqueline Matthews have noted, one of the first things anybody did was compare them to indigenous Australians. Five years later the English biologist Thomas Huxley, a champion of the work of Charles Darwin, described the skulls of Australians as being "wonderfully near" those of the "degraded type of the Neanderthal." It was clear what they were insinuating. If any people on earth were going to have something in common with these now-extinct humans, European scientists assumed, it could only be the strange ones they called savages. Who else could it be but the people who were closest to nature, who had never fit the scientists' definition of what a modern human was?

| | | | | | | | | |

We are forever chasing our origins.

When we can't find what we want in the present, we go back, and back further still, until there, at the dawn of time, we imagine we've found it. In the gloomy mists of the past, having squeezed ourselves back into the womb of humanity, we take a good look. Here it is, we say with satisfaction. Here is the root of our difference.

Once upon a time, scientists were convinced that Aboriginal Australians were further down the evolutionary ladder from other humans, perhaps closer to Neanderthals. In 2010 it turned out that Europeans are actually likely to have the most drops of Neanderthal blood, metaphorically speaking. In January 2014 an international team of leading archaeologists, geneticists, and anthropologists confirmed that humans outside Africa had bred with Neanderthals. Those of European and Asian ancestry have a very small but tangible presence of this now-extinct human in our lineage, up to around 4 percent of our DNA. People in Asia and Australia also bear traces of another known archaic human, the Denisovans. There is likely to have been breeding with other kinds of human as well. Neanderthals and Denisovans, too, mated with each other. Many in the deep past, it seems, were pretty indiscriminate in their sexual partnerships.

"We're more complex than we initially thought," explains John Shea. "We initially thought there was either a lot of interbreeding or no interbreeding, and the truth is between those goalposts somewhere."

The discovery had important consequences. It raked up a controversial, somewhat marginalized scientific theory that had been doing the rounds a few decades earlier. In April 1992 an article had been published in *Scientific American* magazine with the incendiary title "The Multiregional Evolution of Humans." The authors were Alan Thorne, a celebrity Australian anthropologist, who died in 2012, and Milford Wolpoff, a cheery anthropologist based at the University of Michigan, where he still works today. They hypothesized that there was something deeper to human difference, that perhaps we hadn't all come out of Africa as fully modern humans after all.

Although this notion had been mooted before, for Wolpoff, this idea became cemented in the seventies. "I traveled and I looked, I traveled and I looked, I traveled and I looked," he tells me. "And what I noticed was that in different regions, big regions—Europe, China, Australia, that is what I mean by regions, not small places—in different regions, it seemed to me there was a lot of similarity in fossils." That is, they were "similar" in their difference: "They weren't the same and they all were evolving."

His big realization came in 1981 when Wolpoff was working with a fossilized skull from Indonesia slightly to the northwest of Australia, dated at roughly a million years and possibly older. A million years is an order of magnitude older than modern humans, hundreds of thousands of years before some of our ancestors first began to migrate out of Africa. It couldn't possibly be the ancestor of any living person. Yet Wolpoff says he was struck by the similarities he thought he could see between its facial structure and that of modern-day Australians. "I had reconstructed a fossil that looked so much like a native Australian to me I almost dropped it," he says. "I propped it up on my lap with the face staring at me. . . . When I turned it over on its side to get a good look at it, I was really surprised."

Teaming up with Alan Thorne, who had done related research and shared his interpretation of the past, they came up with the theory that *Homo sapiens* evolved not only in Africa, but that some of the earlier ancestors of our species spread out of Africa and then independently evolved into modern humans, before mixing and interbreeding with other human groups to create the one single species we recognize today. In their article for *Scientific American*, which helped catapult their multiregional hypothesis into the mainstream, they wrote, "Some of the features that distinguish major human groups, such as Asians, Australian Aborigines and Europeans, evolved over a long period, roughly where these people are found today."

They described these populations as "types," judiciously steering clear of the word "race." "A race in biology is a subspecies," Wolpoff clarifies when I ask him about it. "It's a part of a species that lives in its own geographic area, that has its own anatomy, its own morphology, and can integrate with other subspecies at the boundaries. . . . There are no subspecies anymore. There may have been [human] subspecies in the past—that's something we argue about. But we do know there are no subspecies now."

Many academics found Wolpoff and Thorne's idea unconvincing or offensive, or both. According to Billy Griffiths, the multiregional way of thinking about human origins, which undercuts the fundamental belief that we are all one species and nothing else, has echoes of an earlier intellectual tradition. "Wherever we are in the world we look at the deep past and these immense spans of time through the lens of our present moment and our biases and what we want," he tells me. "Archaeology is a discipline that is saturated by colonialism, of course. It can't entirely escape its colonial roots." Multiregionalism was a response to the available evidence at the time, but it also suggested that there must be something that profoundly sets "races" apart, that the roots of human difference aren't recent, but actually run deep in time and, consequently, also in our minds and bodies. Its gives rise to the possibility that our origins aren't quite so shared. "That's the ugly political legacy that dogs the multiregional hypothesis," he says.

Wolpoff has always been sensitive to the controversy he helped to stoke. He faced down plenty of criticism when he and Thorne published their work. "We were the enemy," he recalls. "If we were right, there couldn't be a single recent origin for humans. . . . They said, you're talking about the evolution of human races in separate places independently of each other."

And their theory remains unproven. Academics in the West and in Africa today generally accept that humans became modern in Africa and then adapted to the environments where they happened to move to fairly recently in evolutionary time—these are only superficial adaptations, such as skin color. But not everyone everywhere agrees. In China, there's a common belief among both the public and leading academics that Chinese ancestry goes back considerably further than the migration out of Africa. One of Wolpoff's collaborators, Wu Xinzhi, a paleontologist at the Chinese Academy of Sciences, has argued that fossil evidence supports the notion that *Homo sapiens* evolved separately in China from earlier human species who were living there more than a million years ago, despite data showing that modern Chinese populations carry about as much of a genetic contribution from modern humans who left Africa as other non-African populations do.

"There are many people who are not happy with the idea of African origin," I'm told by Eleanor Scerri, an archaeologist based at the University of Oxford who researches human origins. "They have co-opted multiregionalism to make a claim that this is a simplistic idea, that races are real, and that people who have come from a particular area have always been there." She tells me that this thinking appears to be prevalent not only in China but also in Russia. "There is no acceptance that they were ever African."

For some, an unwillingness to accept African origins may be motivated by racism or nationalism, but this isn't the case for all. There are those for whom it's simply a way of squaring old origin stories with modern science. In Australia, for instance, Billy Griffiths tells me, many indigenous people favor the multiregional hypothesis because it sits closer to their own belief that they have been here from the very beginning. Indeed, this is an origin myth shared by cultures in many parts of the world. Until further evidence comes along (and maybe even after it does), the theory of a people's origins can be to some extent a matter of choice, affected as much by personal motivations as by data. The past can never be completely known, so the classic multiregional hypothesis may hang on, despite its lack of support among scientists. It still has power.

While classic multiregionalism seems unlikely to be the story of our past, the fact that we now know our ancestors bred with other kinds of archaic humans does have implications. It gives nourishment to those who would like to resurrect the multiregional hypothesis in full. It's a factual nugget that feeds fresh speculation about the roots of racial difference. Some dogged supporters of the multiregional hypothesis can rightly claim that at least one prediction made by Wolpoff and Thorne has turned out to be correct. The pair suggested that other now-extinct humans such as Neanderthals either evolved into modern humans or interbred with them. And on interbreeding, we now know from genetic evidence, the pair got it right. Some of our ancestors did mate with Neanderthals, although their contribution to our DNA today is so tiny that this couldn't have been particularly widespread. But it did happen.

When I ask Wolpoff if he feels vindicated by this, he laughs. "You said 'vindicated.' We said 'relief'!"

Genetics has done the unthinkable, says the rock-art expert Benjamin Smith. "The thing that has worried me is the way that genetics research has moved. . . . We thought that we were basically all the same, whether you're a bushman in southern Africa, an Aboriginal Australian living in rural Western Australia, or someone like myself who is of European extraction. Everyone was telling us that we were all identical, all the modern science." The latest discoveries appear to move the story a little closer back to the nineteenth-century account. "This idea that some of us are more interbred with Neanderthals, some of us are more interbred with Denisovans . . . and Aboriginal Australians had quite a high proportion of Denisovan genetics, for example. That could lead us back to the nasty conclusion that we are all different," he warns. "I can see how it might be racialized."

Indeed, when the Neanderthal connection was revealed by geneticists, personal ancestry-testing companies were quick to sell services offering paying members of the public the opportunity to find out how much Neanderthal ancestry they have, presumably in the expectation that this might mean something to them. The finding also had a peculiar effect on scientific research. Fairly soon after it was found that it was modern-day Europeans who have the closer association to Neanderthals—not, as it turned out, Aboriginal Australians—the image of the Neanderthal underwent a dramatic makeover. When their remains were first discovered in 1856, the German naturalist Ernst Haeckel had suggested naming them *Homo stupidus*. But now these same Neanderthals, once the dictionary definition of simple-minded, loutish, uncivilized thugs, became oddly rehabilitated.

Svante Pääbo, the director of the genetics department at the Max Planck Institute for Evolutionary Anthropology, in Leipzig, Germany, who spearheaded some of the research that led to the discoveries of ancient interbreeding in the first place, was among those to marshal efforts to compare the genomes of Neanderthals and *Homo sapiens*, in the search for what differs as well as what is present in both. And this was accompanied by plenty of speculation from others. In 2018 a set of researchers in Switzerland and Germany suggested that Neanderthals actually had quite "sophisticated cultural behaviour," prompting one British archaeologist to wonder whether "they were a lot more refined than previously thought." An archaeologist in Spain claimed that modern humans and Neanderthals must have been "cognitively indistinguishable." A few even raised the possibility that Neanderthals could have been capable of symbolic thought, pointing to freshly discovered cave markings in Spain that appear to predate the arrival of modern humans (the finding failed to convince Benjamin Smith).

"Neanderthals are romanticized," John Shea tells me. They're no longer around, and we don't have a great deal of evidence about what they were like or how they lived, which means they can be whatever we want them to be. "We're free to project good qualities, things we admire, and the ideal on them." In reality, whatever they were like, he says, "The interbreeding thing is more like a symbolic thing for us than it is of evolutionary consequence."

Yet researchers haven't been able to help themselves from looking for evolutionary consequences. One team of scientists claimed that the tiny peppering of Neanderthal DNA may have given Europeans different immune systems from Africans. Another published paper linked Neanderthal DNA to a whole host of human differences, including "skin tone and hair color, height, sleeping patterns, mood, and smoking status." An American research group went so far as to try to link the amount of Neanderthal DNA people have with the shape of their brain, implying that non-Africans may have some mental differences from Africans as a result of their interbreeding ancestors.

For more than a century the word "Neanderthal" had been synonymous with low intelligence. In the space of a decade, once the genetic link to modern Europeans was suspected, that all changed. In the popular press, there was a flurry of excitement about our hitherto undervalued relatives. Headlines proclaimed that "we haven't been giving Neanderthals enough credit" (*Popular Science*), that they "were too smart for their own good" (*The Telegraph*), that "humans didn't outsmart the Neanderthals" (*Washington Post*). Meanwhile a piece in the *New Yorker* whimsically reflected on their apparent everyday similarity to humans, including the finding that they may have suffered from psoriasis. Poor things, they even itched like us. "With each new discovery, the distance between them and us seems to narrow," wrote the author. In the popular imagination, the family tree had gained a new member.

In January 2017 the *New York Times* ran a story headlined "Neanderthals Were People, Too" and asked, "Why did science get them so wrong?" This was indeed the big question. If the definition of "people" had always included archaic humans, then why should Neanderthals so suddenly and so generously be accepted as "people" now? And not just accepted, but elevated to the celebrity status of sadly deceased genius cousin? It wasn't all that long ago that scientists had been reluctant to accept the full humanity even of Aboriginal Australians. Gail Beck's family had been denied their culture; treated in their own nation as unworthy of survival; their children ripped from their parents to be abused by strangers. In the nineteenth century Aboriginal Australians had been lumped together with Neanderthals as evolutionary dead ends, both destined for extinction. But now that common ground had been found between Europeans and Neanderthals, *now* we were all people! *Now* we had found our common ground!

If it had turned out that Aboriginal Australians were the ones to possess that tiny bit of Neanderthal ancestry instead of white people of European descent, would our Neanderthal cousins have found themselves quite so remarkably reformed? Would they have been welcomed with such warm hugs? It's hard not to see the public and scientific acceptance of Neanderthals as "people like us" as another manifestation of the Enlightenment habit of casting humanity in the European image. In this case Neanderthals have been drawn into the circle of humankind by virtue of being just a little related to Europeans—forgetting that a century ago, it was their supposed resemblance to indigenous Australians that helped cast the latter, actual living human beings, out of the circle.

| | | | | | | | | |

Milford Wolpoff is clear with me that he doesn't think there is any biological basis to race, that there are no separate races, except as social categories. He comes across as honest and well meaning, and I believe him. The more we speak, the more I like him. But one obvious implication of his multiregional hypothesis is that if different populations became modern in their own way on their own territories, then maybe some became what we today recognize as human sooner than others. "A modern human from China looks different than a modern human in Europe, not in the important ways, but in other ways," he tells me. "So did one become modern earlier than the other one?" It's a line of thinking that opens a door for the politics of today to be projected onto the past, that gives rise to racial speculation even if that's not what he intends.

There is still not enough evidence that any humans became modern outside of Africa in the way that classic multiregional theory suggests. Even Wolpoff concedes that Africa must remain at the heart of the story. "I will never say that all of modernity is African, but you've gotta think that most of it is," he tells me, if only because in our deep past that's where most people lived. It is impossible to airbrush Africa out of the lineage of every living person. The genetic evidence we have to date confirms that some version of an "out of Africa" scenario must have happened.

But over time, the picture inside Africa has changed to incorporate the growing scientific realization that our origins might have been a little fuzzier than we imagine. In the summer of 2018 Eleanor Scerri, at the Institute of Archaeology in Oxford, together with a large international team of geneticists and anthropologists, published a scientific paper suggesting that rather than humans evolving from a single lineage that can be traced to a single small sub-Saharan African population, perhaps our ancestors were the product of many populations across a far wider area within Africa. These Pan-African populations might have been isolated by distance or ecological barriers, and could therefore have been very different from one another. It is multiregionalism, if you like, but within one continent.

"Gradually we started to emerge from the occasional mixing of the populations that were spread around," Scerri tells me. "The characteristics that define us as a species don't appear in any single individual until much later. Before that, the characteristics of our species were distributed across the continent in different places at different times." Modern humans, *Homo sapiens*, emerged from this "mosaic." "We need to look at all of Africa to get a good picture of origins." This version of our past still puts Africa at its center, as the first home of our ancestors, but it also concedes that modern humans didn't appear suddenly in one place looking and sounding sophisticated and producing representational art. There was no sudden moment at which the first modern human emerged. The characteristics of humans existed in various others before us.

John Shea agrees: "Humans evolved in Africa first. Not in just one garden of Eden, but among a broadly distributed population more or less like stops across a subway system. People were moving around along the rivers and coastlines." In short, according to this view we are a product of longer periods of time and space, a mixture of qualities that incubated in Africa.

According to Martin Porr, the archaeologist in Australia, this version of the past is more plausible given the way that fossil evidence is scattered across the African continent. For him personally, it also resonates with indigenous Australian ways of defining what it means to be human. Up north in the Kimberley where he has worked most of the time, he says, rock art is not thought of as just images upon rock. "The rock is actually not a rock but it's a formation out of the dreamtime that is alive, that is in the living world, that people inhabit. And people themselves are part of that." Human and object, object and environment, are not separated by hard divisions the way they are in most Western worldviews.

"You can oscillate in and out of humanity just as objects and animals can oscillate between being human." An inanimate object can take on human qualities, the way a doll does to a child. In that sense, too, Porr suggests that what made a being human in the past also oscillated.

"I think there's nothing essential about human beings at all." This, he explains, is how he has come to think about our origins. Not that our evolutionary journey was one big leap, but that we are the gradual products of elements that already existed, not only in our African ancestors but also in Neanderthals, Denisovans, and other archaic humans. Perhaps some of what we think of as purely human characteristics exist in other living creatures today, too.

It's a radically different way of thinking about what it means to be human—ditching the European Enlightenment view and taking a cue instead from other cultures and older systems of thought. It's a challenge to researchers who have dedicated their careers to identifying the first modern humans and defining what they were like, chasing the tail of the European Enlightenment philosophers who thought they already knew. Archaeologists are still trying to hunt down the earliest cave art, the earliest sign of symbolic thought, in the hope of pinpointing the magic moment at which we emerged, and where. Geneticists, too, hunt for magical ingredients in our genome, the ones that will indicate what makes us so remarkable. Yet increasingly the evidence suggests that it was never that simple.

"Very few people like looking at human origins from a postcolonial context, but there is a broader story," says Porr. There are other ways of picturing humanity than as a uniquely special entity far removed from all other living things. Eleanor Scerri agrees that fresh scientific findings are forcing a rethink of what it means to be human. "Popular science needs to get away from this idea that we originated, and that was *us*. There's never a time that we were not changing," she says, "the idea of these immutable forms, and that we originate in one place and that's who we are, that's where we're from."

What does this mean for us today? If we can't agree on what makes a modern human, then where does that leave the idea of universal humanity?

If our origins aren't crystal clear then how do we know that we're all the same? What does it mean for race?

In a sense, it shouldn't matter. How we choose to live and treat each other is a political and ethical matter, one that's already been decided by the fact that as a society we have chosen to call ourselves human and give every individual human rights. In reality, though, the political tentacles of race reach into our minds and demand proof. If we are equally human, equally capable and equally modern, then there are those who need convincing before they grant full rights, freedoms, and opportunities to those they have historically treated as lesser. They need to be convinced before they will commit to redressing the wrongs of the past, before they agree to affirmative action or decolonization, before they fully dismantle the structures of race and racism. They're not about to give away their power.

And if we're honest, maybe we all need to be convinced. Many of us hold subtle prejudices, unconscious biases, and even just popular beliefs that betray our suspicion that we're not quite the same. We cling to race even when we know we shouldn't. A British friend of mine, of mixed Pakistani and white English ancestry, who has never been to Pakistan and has no real ties to the country anymore, told me recently that she believed there is something in her blood, something biologically deep within her that makes her feel Pakistani. I feel this way sometimes about my Indian heritage, although in my case I have strong cultural ties to the country of my parents. But where does culture end and ethnicity begin? Those of us who cherish our ethnic or racial identities, perhaps we, too, whether on the political left, right, or center, have some commitment to the idea of racial difference. Whether it is those supporters of President Donald Trump who speak of white supremacy, or whether it is Senator Elizabeth Warren, who apparently felt the need to prove her Native American ancestry through a DNA test when it was questioned by Trump, people everywhere of all political persuasions find it easy to buy in to the idea that biological race means something.

This is a problem for science. When Enlightenment thinkers looked at the world around them, some took the politics of their day as the starting point. It was the lens through which they viewed all human difference. We do the same today. The facts only temper what we think we already know. Even when we study human origins, we don't actually start at the beginning. We start at the end: with our assumptions as the basis for inquiry. We need to be persuaded before we cast aside our prior beliefs about who we are. The way new research is interpreted is always at the mercy of the old ideas.

"You can either use the present to explain the past. Or you can use the past to explain the present," John Shea tells me. "But you can't do both." To make sense of the past—and of ourselves—is not a simple job of gathering scientific data until we have the truth. It isn't just about how many fossils we have or how much genetic evidence. It's also about squaring the stories we have about who we are with the information we're given. Sometimes this information becomes slotted into the old stories, reinforcing them and giving them strength, even if it needs to be forced like a round peg into a square hole. Other times we have to face the uncomfortable realization that a story must be ditched and rewritten because however hard we try it no longer makes sense.

But the stories we're raised on—the tales, myths, legends, beliefs, even the old scientific orthodoxies—are how we frame everything we learn. The stories are our culture. They are the minds we inhabit. And that's where we have to start.

CHAPTER 2

# It's a Small World

*How did scientists enter the story of race?*

Once, a long time ago, I floated around the earth in the space of minutes.

I was on a ride at the Magic Kingdom in Walt Disney World, in Florida; my little sisters and I were perched alongside each other in a slow mechanical boat, buoyed by sugar. "It's a Small World (After All)" chimed in tinny children's voices, while minuscule automata played out cultural stereotypes from different countries. From what I can recall, there were spinning Mexicans in sombreros and a ring of African dancers laughing alongside jungle animals. Indian dolls rocked their heads from side to side in front of the Taj Mahal. We sailed past, given just enough time to recognize each cultural stereotype, but not quite enough time to take offense.

This long-forgotten vignette from my childhood is what comes back to me on the drizzly day I approach the eastern corner of the Bois de Vincennes woodland in Paris. I had heard that somewhere here I would find the ruins of a set of enclosures in which humans were once kept—not as cruel punishment by the authorities, and not by some murderous psychopath. Apparently they were just ordinary, everyday people, kept here by everyday people, for the fascination of millions of other everyday people, for no other reason than where they happened to come from and what they happened to look like.

"Man is an animal suspended in webs of significance he himself has spun," the anthropologist Clifford Geertz wrote in 1973. These webs are ours only until someone comes along to pull at the threads. The nineteenth century was an age marked by unprecedented movement and cultural contact that turned the world into a smaller place than it had ever been. It was less mysterious, perhaps, but no less fascinating. And people wanted to see it all. So in 1907 the grand Colonial Exposition was held at an overgrown site in Paris within the Bois, the Garden for Tropical Agriculture, recreating different parts of the world in which France had its colonies.

Eight years earlier, in 1899, the garden had been founded as a scientific project to see how crops might be better cultivated in the tropics and thus help to bring in more income for colonizers back in Europe. This exposition went a step further. To exotic plants and flowers it added people, displaying them in houses vaguely typical of the ones they might have left behind, or at least what Europeans imagined them to be. There were five mini "villages" in all, each designed to be as realistic as possible so visitors could experience what normal life was like for these foreigners. It was an Edwardian Disneyland, not with little dolls but with actual people.

They transformed the tropical garden into nothing less than a human zoo.

"In Paris, there were many exhibitions with human zoos," I'm told by the French anthropologist Gilles Boëtsch, former president of the scientific council at the National Center for Scientific Research, who has studied their dark history. There was a circus element to it all; it was a cultural extravaganza. But there was also a genuine desire to showcase human diversity, to give a glimpse of life in the faraway colonies. According to some estimates, the 1907 Paris exposition attracted two million visitors in the space of just six months—it was a hit with curious citizens who wanted to see the world in their backyard.

Wherever they were held, most evidence of the human zoos has long disappeared, most likely deliberately forgotten. The Garden for Tropical Agriculture is one rare exception. That said, the French authorities don't appear to want to brag about it. It's tucked behind some quiet and well-to-do apartment blocks with barely any signposting. Greeting me as I enter is a Chinese arch that was once probably bright red, but has since faded to a dusty gray. As I walk under it, down a gravel path, the place is peaceful but dilapidated. To my surprise, most of the buildings have survived the last century fairly intact, as though everything was abandoned immediately after the visitors left.

To one side is a weathered sculpture of a naked woman, reclining and covered in beads, her head gone, if it was ever there at all. A solitary jogger runs past.

For European scientists, zoos like this offered more than fleeting amusement value. They were a source of biological data, a laboratory stocked with captive human guinea pigs. "They came to the human zoos to learn about the world," explains Boëtsch. Escaping the bother of long sea voyages to the tropics, anatomists and anthropologists could conveniently pop down to their local colonial exhibition and sample from a selection of cultures in one place. Researchers measured head size, height, weight, and color of skin and eyes, and recorded the food the subjects ate, documenting these observations in dozens of scientific articles. With their notes they set the parameters for modern race science.

Race was a fairly new idea. Some of the first known uses of the word date from as recently as the sixteenth century, but not in the way we use it now. Instead, at that time it referred to a group of people from common stock, like a family, a tribe, or perhaps—at a long stretch—a small nation. Even up until the European Enlightenment in the eighteenth century, many still thought about physical difference as a permeable, shifting quantity. It was rooted in geography, perhaps explaining why people in hotter regions had darker skins. If those same people happened to move somewhere colder, it was assumed their skins would automatically lighten. People could shift their identity by moving away or converting to another religion.

The notion that race was a hard and fixed feature that people couldn't choose, an essence passed down to their children, came slowly, in large part from Enlightenment science. The eighteenth-century Swedish botanist Carl Linnaeus, famous for classifying the natural world from the tiniest insects to the biggest beasts, turned his eye on humans. If flowers could vary by color and shape, then perhaps humans could also be classified into groups. In the tenth edition of *Systema Naturae*, a catalogue he published in 1758, Linnaeus laid out the categories we still use today. He listed four main categories of human, corresponding to the Americas, Europe, Asia, and Africa, respectively, and each easy to spot by their color: red, white, yellow, and black.

Categorizing humans became a never-ending business. Every gentleman scholar (and they were almost exclusively men) drew up his own dividing lines, some going with as few as a couple of races, others with dozens or more. Many never saw the people they were describing, instead relying on secondhand accounts from travelers, or just hearsay. Linnaeus himself included two separate subcategories within his *Systema Naturae*, monsterlike humans and feral humans. But once defined, these "races" rapidly became slotted into hierarchies based on the politics of the time, character being conflated with appearance, and political circumstance becoming a biological fact. For example, Linnaeus described indigenous Americans, his "red" race, as having straight black hair and wide nostrils, but also as being "subjugated," as though subjugation were in their nature.

And so it began. By the time human zoos were a popular attraction—when the ghostly enclosures I'm walking around in the Bois de Vincennes were not eerily empty as they are now but full of performers, when I more likely would have been inside a cage than outside it—the parameters of human difference had become hardened into those we recognize today.

Paris wasn't the only city to enjoy this type of spectacle, of course. Other European colonial powers hosted similar events. Indeed by the time of the 1907 Paris exposition, human zoos had been around for more than a century. In 1853 a troupe of Zulus made a grand tour of Europe. Even earlier, an advertisement in London's *Morning Post* newspaper of September 20, 1810, signaled the arrival of one of the most notorious of all racial freak shows. "From the Banks of the River Gamtoos, on the Borders of Kaffraria, in the interior of South Africa, a most correct and perfect Specimen of that race of people," it announced. The "Hottentot Venus," as she was described, was available for anyone to take a peek, for a limited time only and at the cost of two shillings. Her real name was Saartjie Baartman, and she was aged somewhere between twenty and thirty. What made her so fascinating were her enormous buttocks and elongated labia, considered by Europeans to be sexually grotesque. Calling her a "Venus" was a joke at her expense.

The *Morning Post* took pains to mention the expense shouldered by a Boer farmer, Hendric Cezar, in transporting Baartman all the way to Europe. He was banking on her body, which caused a scandal. She had been Cezar's servant in Africa, and by all accounts she had come with him to Europe of her own free will. But it's unlikely that the life she endured as his traveling exhibit was what she expected. Her career was brief and humiliating. At each show she was brought out of a cage to parade in front of visitors, who poked and pinched to check that she was real. Commentators in the press couldn't help but notice how unhappy she seemed, even remarking that if she felt ill or unwilling to perform, she was physically threatened. To add to the humiliation, she became, quite literally, the butt of jokes across the city, rendered in relentless caricature.

At the end of her run, Baartman ended up in Paris. She fell at the mercy of a celebrated French naturalist, Georges Cuvier, who was a pioneer in the field of comparative anatomy, which aims to understand the physical differences between species. Like so many before him, he was spellbound by her—but his was an anatomist's fascination, one that drove him to undertake a detailed study of every bit of her body. When she died in 1815, just five years after being displayed in London, Cuvier dissected her, removing her brain and genitals, and presented them in jars to the French Academy of Sciences.

As far as Cuvier was concerned, this was just science and she was just another sample. The prodding, cutting, dehumanizing fingers of researchers like Cuvier sought only to understand what made her and those like her different. What gave some of us darker skin and others light? Why did we have different hair, body shape, habits, and language? If we were all one species, then why didn't we look and behave the same? These were questions that had been asked for decades already, but it was nineteenth-century scientists who really turned the study of humans into the most gruesome art. People became objects, grouped together like museum exhibits. Any sense of common humanity was left at the door, replaced by the cold, hard tools of dissection and categorization.

Following a lifetime of being relentlessly poked and prodded, Baartman continued to be on show for a hundred and fifty years after her death. Her abused body ended up at the Musée de l'Homme, the Museum of Man, near the Eiffel Tower; a plaster cast of it stood there until as recently as 1982. It was only in 2002, after a request from Nelson Mandela, that her remains were removed from Paris and finally returned to South Africa for burial.

| | | | | | | | | |

"In the modern world we look to science as a rationalization of political ideas," I'm told by Jonathan Marks, a genial, generous professor of anthropology at the University of North Carolina, Charlotte. He is one of the most outspoken voices against scientific racism. Race science, he explains, emerged "in the context of colonial political ideologies, of oppression and exploitation. It was a need to classify people, make them as homogeneous as possible." Grouping people made it easier to control them.

It is no accident that modern ideas of race were formed during the height of European colonialism, when those in power had already decided on their own superiority. By the nineteenth century, the possibility that races existed and some were inferior to others gave colonialism a moral kick in the drive for public support. The truth—that European nations were motivated by economic greed or power—was harder to swallow than the suggestion that the places they were colonizing were too uncivilized and barbaric to matter, or that they were actually doing the savages a favor.

In the United States, the same tortured logic was used to justify slavery. The transatlantic trade in slaves officially ended when the United Kingdom passed the Slave Trade Act in 1807; the United States banned the import of slaves the following year. But within the United States the trade continued, and the use of slave labor wouldn't stop for many decades more. Black bodies were plundered both in life and in death. For instance, the corpses of black slaves were routinely stolen or sold for medical dissection. Daina Ramey Berry, a history professor at the University of Texas at Austin, who has documented the economic value of slavery in the United States, notes that there was a brisk trade in dead black people in the nineteenth century, some of which were exhumed by their owners for a quick profit. It's ironic that much of our modern scientific understanding of human anatomy was built on the bodies of those who were considered at the time less than human.

"If you could say that the slavers were naturally distinct from the slaves, then you have essentially a moral argument in favor of slavery," says Marks. Given this distinction, many feared that the abolition of slavery would set free the human zoo, unleashing chaos. In 1822 a group calling itself the American Colonization Society bought land in West Africa to establish a colony named Liberia, now the Republic of Liberia, motivated largely by the desperate dread that freed black slaves would want to settle among whites, with the same rights. Repatriation to the continent of their ancestors seemed like a convenient solution, ignoring that after generations in slavery, most black Americans simply didn't have a tangible connection to Africa anymore—let alone to a new country that their ancestors most likely had never seen.

Louis Agassiz, a Swiss naturalist who had been mentored by Georges Cuvier and moved to the United States in 1846, argued passionately against blacks being treated the same as whites. Shaken by such an intense physical disgust toward black domestic workers serving him food at a hotel that he almost couldn't eat there at all, he became convinced that separate races originated in different places, and possessed different characters and intellectual abilities.

Responsibility for being enslaved was turned back on the slaves themselves. They were in this miserable, degrading position not because they had been forcibly enslaved, it was argued, but because it was their biological place in the universe. At a meeting of the British Association for the Advancement of Science in Plymouth in 1841, an American slave owner from Kentucky, Charles Caldwell, had already claimed that Africans bore more of a resemblance to apes than to humans. In their 1854 book *Types of Mankind*, an American physician, Josiah Clark Nott, and an Egyptologist, George Gliddon, went so far as to sketch actual comparisons between the skulls of white and black people alongside those of apes. While the typical European face that they drew was artfully modeled on classical sculpture, African faces were crude cartoons, with exaggerated features that made it seem they had more in common with chimpanzees and gorillas than with humans.

Propelled by a belief that black people had their own unique diseases, Samuel Cartwright, a medical doctor practicing in Louisiana and Mississippi, in 1851 characterized what he saw as a mental condition particular to black slaves, which he coined "drapetomania," or "the disease causing Negroes to run away." Harvard University historian Evelynn Hammonds, who teaches Cartwright's story to her students, laughs darkly when she recounts it: "It makes sense to him, because if the natural state of the negro is to be a slave, then running away is going against their natural state. And therefore it's a disease."

For Hammonds, the chilling aspect of Cartwright's work is the way in which he methodically described the symptoms of drapetomania. "'The color of the skin is the main difference,'" she reads for me from her notes, "'the membranes, the muscles, the tendons, all fluids and secretions, then the nerves, and the bile. There's a difference in the flesh. The bones are whiter and harder, the neck is shorter and more oblique.'" Cartwright continues this way, couching racism in medical terminology. "These kinds of observations turned into questions to be explored going forward," says Hammonds. "Since the 1850s, people have been trying to figure out if black bones are harder than white bones." Cartwright's medical "discoveries" were patently rooted in the desire to keep slaves enslaved, to maintain the status quo in the American South, where he lived. In place of universal humanity came a self-serving version of the human story in which racial difference became an excuse for treating people differently. Time and again, science provided the intellectual authority for racism, just as it had helped define race to begin with.

Race science became a pastime for nonscientists, too. The French aristocrat and writer Count Arthur de Gobineau, in *An Essay on the Inequality of the Human Races*, published in 1853, proposed that there were three races, in what he saw as an obvious hierarchical relationship. "The negroid variety is the lowest, and stands at the foot of the ladder. . . . His intellect will always move within a very narrow circle," he stated. Pointing to the "triangular" face shape of the "yellow race," he explained that this was the opposite of the negroid variety. "The yellow man has little physical energy, and is inclined to apathy. . . . He tends to mediocrity in everything."

Neither could be a match for his own race.

Reaching his predictable pinnacle, Gobineau wrote, "We come now to the white peoples. These are gifted with reflective energy, or rather with an energetic intelligence. They have a feeling for utility, but in a sense far wider and higher, more courageous and ideal, than the yellow races." His work was a naked attempt to justify why those like him deserved the power and wealth they already had. This was the natural order of things, he argued. He didn't need hard evidence for his theories because there were plenty of people around him ready and willing to agree that they, too, belonged to a superior race.

Later, Gobineau's ideas helped reinforce the myth of racial purity and the creed of white supremacy. "If the three great types had remained strictly separate, the supremacy would no doubt have always been in the hands of the finest of the white races, and the yellow and black varieties would have crawled forever at the feet of the lowest of the whites," he wrote, promoting a notion of an imaginary "Aryan" race. These glorious Aryans, he believed, had existed in India many centuries ago, had spoken an ancestral Indo-European language, and had since spread across regions of the world, diluting their superior bloodline.

Myth and science coexisted, and both served politics. In the run-up to the passage in 1865 of the Thirteenth Amendment, which abolished slavery in the United States, the race question wasn't resolved—it just became thornier. Although many Americans believed in emancipation on moral grounds, fewer were convinced that full equality would ever be possible, for the simple reason that groups weren't biologically the same. Even Presidents Thomas Jefferson and Abraham Lincoln believed that blacks were inherently inferior to whites. Jefferson, himself a slaveholder, agreed with those who thought that the solution to the problem of what to do with freed slaves was to send them to a colony of their own. Freedom was framed as a gift bestowed on unfortunate black slaves by morally superior white leaders, rather than as a reflection of a hope that everyone would one day live alongside one another as friends, colleagues, and partners.

| | | | | | | | | |

Not all scientists, of course, were quite so self-serving. For those who wanted to establish the facts about human difference, there were unanswered questions. The biggest puzzle was that there was no fleshed-out mechanism for how different races—if they were real—might have emerged. If each race was distinct, then where did they each come from, and why? Going by the Bible, as many Europeans did, one explanation for the different races was that, after the big flood, Noah's children spread to different parts of the Earth.

In 1871 the naturalist and biologist Charles Darwin published *The Descent of Man*, sweeping away these religious creation myths and framing the human species as having one common ancestor many millennia ago and as having evolved slowly like all other life on earth. After studying humans across the world, their emotions and expressions, he wrote, "It seems improbable to me in the highest degree that so much similarity, or rather identity of structure, could have been acquired by independent means." We are too alike in our basic responses, our smiles and tears, our blushes, to have different origins. On this point alone Darwin might have settled the race debate. He demonstrated that we could only have evolved from shared origins, that human races didn't emerge separately.

And on a personal level, this was important to him. Darwin's family included two influential abolitionists, his grandfathers, Erasmus Darwin and Josiah Wedgwood. He himself had seen the brutality of slavery firsthand on his travels. When the naturalist Louis Agassiz, in the United States, spoke about human races as having separate origins, Darwin wrote disparagingly in a letter that this must have come as comfort to slaveholding Southerners. A central tenet of the antislavery movement was that humanity is one, that we share the same blood.

But the abolitionists' conviction wasn't the last word on the subject. Darwin still struggled with the notion of equality when it came to race. Like Abraham Lincoln, who was born on the same day as Darwin, he opposed slavery but was also ambiguous on the question of whether black Africans and Australians were strictly equal to white Europeans. He left open the possibility that, even though we could all be traced back to a common ancestor, populations may have diverged since then, producing levels of difference. As the British anthropologist Tim Ingold notes, Darwin saw gradations between the "highest men of the highest races and the lowest savages." He suggested, for example, that the "children of savages" have a stronger tendency to protrude their lips when they sulk than European children, because, similar to chimps, they are closer to the "primordial condition." The historian Gregory Radick at the University of Leeds notes that Darwin, even though he made such a bold and original contribution to the idea of racial unity, also seemed to be unembarrassed by his belief in an evolutionary hierarchy. Men were above women, and white races were above others.

Mixed with the politics of the day, this was devastating. The uncertainty around the biological facts left more than enough room for ideology to be mixed with real science, leading to the fabrication of fresh racial myths. Brown and yellow races were a bit higher up, some argued, not unlike Darwin did. Whites were the most evolved—and by implication, the most civilized and the most human. What was seen as the success of the white races became couched in the language of the "survival of the fittest," which carried the implication that the most "primitive" peoples, as they were described, would inevitably lose the struggle for survival as the human race evolved. Ingold argues that even Darwin himself began to frame evolution as an "imperialist doctrine of progress," rather than seeing it as acting to make a species better adapted to its particular environment.

"In bringing the rise of science and civilisation within the compass of the same evolutionary process that had made humans out of apes, and apes out of creatures lower in the scale, Darwin was forced to attribute what he saw as the ascendancy of reason to hereditary endowment," writes Ingold. "For the theory to work, there had to be significant differences in such endowment between 'tribes' [and] 'nations.'" For hunter-gatherers to live so differently from city dwellers, the logic goes, it must be that their brains had not yet progressed to the same stage of evolution.

Adding fuel to this bonfire of flawed thinking (after all, we now know that the brains of hunter-gatherers are no different from those of anyone else) were Darwin's supporters, some of whom happened to be fervent racists. The English biologist Thomas Henry Huxley, known as "Darwin's bulldog," argued that not all humans were equal. In an 1865 essay on the emancipation of black slaves, he wrote that the average white was "bigger brained," asserting, "The highest places in the hierarchy of civilization will assuredly not be within the reach of our dusky cousins." For Huxley, freeing slaves was a morally good thing for white men to do, but the raw facts of biology made the idea of equal rights—for women as well as for black people—little more than an "illogical delusion." In Germany, too, Darwin's loudest cheerleader was Ernst Haeckel, who taught zoology at the University of Jena beginning in 1862 and was a proud nationalist. He liked to draw connections between black Africans and primates, seeing them as a kind of living "missing link" in the evolutionary chain that connected apes to Europeans.

Darwinism did nothing to slow racism. Instead, ideas about the existence of different races and their relative superiority just became repackaged in new theories. Science—or the lack of it—in the end legitimized racism, rather than quashing it. Whatever real and worthwhile questions might have been asked about human difference were unavoidably tainted by politics and economics.

| | | | | | | | | |

I pick my way through a high thicket of bamboo and find an intricate wooden pagoda.

Farther still in the sunlit Garden for Tropical Agriculture is a Tunisian house, coated in thick green moss. If their histories were unknown to me, I might find the buildings in this quiet maze beautiful. They are grand and otherworldly, ethereal relics of foreign places as imagined by another age. But of course I'm acutely aware that each was also once a kind of home to real people like me, pulled from their lives thousands of miles away for the entertainment of paying visitors. As a reminder, through the smashed window of a Moroccan castle, complete with battlements and blue tiles, I'm caught off-guard by a glaring red face that must have been painted by vandals.

However beautiful they are, these aren't homes at all. They're gilded cages.

It's hard to imagine what life would have been like on the inside of the human zoos, looking out. The people kept here weren't slaves. They were paid, similar to actors under contract, but were expected to dance, act, and carry out their everyday routines in public view. Their lives were live entertainment. Little effort was made to make the people feel comfortable in their temporary homes, much less to acclimatize them. After all, the whole point of this spectacle was to underscore just how different they were, to imagine that even in a cold climate they would choose to walk around in as few clothes as they wore in a hot one, that their behavior couldn't change, no matter where they lived. Visitors were made to believe that the cultural differences were woven into their bodies like stripes on a zebra. "When there was a birth, it meant a new show," Gilles Boëtsch from the National Center for Scientific Research tells me. People would flock to see the baby.

Science had created a distance between the viewers and the viewed, the colonizers and the colonized, the powerful and powerless. For those confronted with people from foreign lands in this way, bizarrely out of context, referenced in a book, or transplanted in some fake village in Paris, it only helped reinforce the notion that we were not all quite the same. As visitors peered into their homes, the performers in human zoos must have been curiosities not just because they looked and behaved differently but also because control of their lives belonged to others who didn't look like them. The ones on the outside of the cage were clothed, civilized, and respectable, whereas those on the inside were seminaked, barbaric, and subjugated.

"People are more readily perceived as inferior by nature when they are already seen as oppressed," write the American scholars Karen Fields and Barbara Fields in their 2012 book *Racecraft*. They explain how a sense of inevitability gets attached to a social routine until it comes to be seen as natural. The idea of race didn't make people treat other people as subhuman. They were already treated as subhuman before race was invoked. But once it was invoked, the subjugation took on a new force.

Something about treating human difference as a science gave it a peculiar quality. Observing humans turned the humans being observed into strange beasts. While maintaining the unimpeachable impression of scientific objectivity, somehow the scientist himself always turned out to be the gold standard of beauty and intelligence. His own race was safe in his hands. The German naturalist Johann Blumenbach idealized the Caucasian race, to which he belonged, but described Ethiopians as being "bandy-legged." If legs were different, there was never any question that it was Caucasians who might be the unusual ones. The creatures caged in the human zoos were those who had failed to reach the ideal of white European physical and mental perfection.

The scientific distance created by believing that racial hierarchies existed in nature, this uneven balance of power, allowed human zoos to treat their performers as less than equals, making life for them fatally precarious. According to Boëtsch, many who lived in the zoos died from pneumonia or tuberculosis. Concerns were expressed in the press. There were always protests, as there had been about Saartjie Baartman, but they made little difference.

In another example, around the same time that the Paris exposition was held, a Congolese "pygmy" named Ota Benga, who had been brought to the United States to be displayed at the St. Louis World's Fair, was installed in the Monkey House at the Bronx Zoo in New York, without shoes. Visitors loved him. "Some of them poked him in the ribs, others tripped him up, all laughed at him," the *New York Times* reported. He was eventually rescued by African American ministers, who found him a place in an orphanage. Ten years later, in despair because he couldn't return home to the Congo, Benga borrowed a revolver and shot himself through the heart.

As I stand among the weeds and crumbling former homes of Paris's human zoo, it's difficult not to see that the reason anyone pursued the scientific idea of race was not so much to understand the differences in our bodies as it was to try to justify why we lead such different lives. Why else? Why would something as superficial as skin color or body shape matter otherwise? What the people who created the human zoos really wanted to know was why some people are enslaved and others free, why some prosper while others are poor, and why some civilizations have thrived while others haven't. Imagining themselves to be looking objectively at human variation, in fact scientists were often looking for answers in our bodies to questions that existed far outside them. Race science had always sat at the intersection of science and politics, of science and economics. Race wasn't just a tool for classifying physical difference but was also a way of measuring human progress, of placing judgement on the capacities and rights of others.

CHAPTER 3

# Scientific Priestcraft

*Deciding that races could be improved, scientists looked for ways to improve their own*

The past is built of the things we choose to remember.

The Max Planck Society, with its administrative headquarters in Munich, Germany, has an illustrious history. It has produced eighteen Nobel Prize winners, including the theoretical physicist Max Planck, after whom it's named. Its institutes employ more than 14,000 scientists, have an annual budget of 1.8 billion euros, and produce more than 15,000 published scientific papers a year. By any standards it's one of the most prestigious science organizations in the world. But in 1997 the biologist Hubert Markl, then president of the society, made a decision that would threaten the reputation of his entire establishment. He wanted to scratch beneath its glorious history to reveal a secret that had been hidden for fifty years.

Before 1948 the Max Planck Society existed in an earlier incarnation, the Kaiser Wilhelm Society for the Advancement of Science. Established in the German Empire in 1911, it was as important then as the Max Planck Society is now, cementing Germany's place in modern scientific history. Albert Einstein did some of his research at one of its institutes, as did other geniuses of the time. But it was later, as the Nazis took power and began to set in motion their own scientific priorities, that things took a disturbing turn.

Whatever had gone on was quietly forgotten after the Second World War, although of course there were always rumors. Researchers were involved in scientific operations under the Third Reich, the whispers suggested, maybe even were party to murder and torture. They must have been. Under the Nazi regime, notes the writer James Hawes, half of the nation's doctors were Nazi Party members. For a decade, German universities taught racial theory.

There was undoubtedly a story to be uncovered, but it was thought wiser to leave it alone. By the Max Planck Society's own admission, there was a tradition of glossing over its ignominious past in favor of celebrating its greater scientific achievements. By the 1990s, though, there was too much pressure from the public to ignore it any longer. And anyway, older members of staff who had been alive during the war—who might be affected by such revelations—had almost all died. The time had come. So Markl resolved to lift the lid and appointed an independent committee to investigate what German scientists at the Kaiser Wilhelm Society might have done during the war.

It would be a survey into the very darkest corners of race science.

It was already clear that figures from within science and academia must have played a role in developing Adolf Hitler's ideology of racial hygiene, which argued that those of pure, "Aryan" racial stock should be encouraged to breed, while others were gradually eliminated—an ideology that culminated in the Holocaust. In hindsight, it couldn't have been done without scientists, both to provide the theoretical framework for such an audacious experiment and to carry out the job itself. On the practical side, there would have been those setting up concentration camps and gas chambers, as well as determining who should die. And then there were all the gruesome human experiments known to have been carried out on people who eventually were killed, and plundering them for biological data. Younger researchers at the Max Planck Society worried that the body of scientific work they had inherited might bear bloody stains.

They were right to worry. The past turned out to be dripping with blood. Within a few years of Markl's launch of the investigation, historians began publishing their findings, and they were devastating. Some had assumed that the Nazis were ignorant of or hostile towards science. Historical research proved this wasn't true. The Kaiser Wilhelm Society's scientists had willingly cooperated with the Nazi state, marrying academic interests and political expediency, helping to secure financial support and social standing for themselves. "Such research not only literally built on the spoils of war, it also led scientists deep into the abysses of Nazi crimes," wrote a reviewer. At least one prominent scientist helped draft and disseminate the legislation around racial ideology.

Those who weren't opportunistic were often complicit, displaying moral indifference even when they saw inhumane or criminal acts taking place right in front of them. When moves began in 1933 to expel Jewish scientists from the Kaiser Wilhelm Society (Einstein abandoned Germany that same year, leaving for a conference and wisely never returning), staff made little effort to stand in the way. At least two of the society's scientists and two other staff members ended up dying in concentration camps.

And then there were those who wholeheartedly supported the Nazis from the beginning. The work of Otmar von Verschuer, head of department at one of the Kaiser Wilhelm Society's institutes, makes for chilling reading. Until the war von Verschuer was a widely respected academic; his research on twins as a way of understanding genetic inheritance had been funded for a few years by the Rockefeller Foundation, in New York. He was even invited to speak at the Royal Society in London. But he was also, it transpired, an anti-Semite who openly praised Hitler and believed in a biological solution to what he saw as Jews' threat to racial purity. According to the American anthropologist Robert Wald Sussman, von Verschuer became one of the Nazis' race experts when it came to addressing the "Jewish question," actively legitimizing the regime's racial policies. One of his former students, the doctor Josef Mengele, went on to become infamous for his cruel experiments on twins and pregnant women at Auschwitz concentration camp. Marek Kohn, a British writer, documented in *The Race Gallery* (1995) that among the samples that von Verschuer had sent to him from Auschwitz were "pairs of eyes from twins . . . dissected after their murder . . . children's internal organs, corpses and the skeletons of murdered Jews."

In 2001 the Max Planck Society at last accepted responsibility for historic crimes committed by its scientists. In its apology the society admitted, "Today it is safe to say that von Verschuer knew of the crimes being committed in Auschwitz and that he, together with some of his employees and colleagues, used them for his purposes." Markl added in his speech, "The Kaiser Wilhelm Society tolerated or even supported research among its ranks that cannot be justified on any ethical or moral grounds. . . . I would like to apologize for the suffering of the victims of these crimes—the dead as well as the survivors—done in the name of science."

This came too late for justice, of course. Those involved had died already. What was remarkable was that it had taken so long to root out the facts of what happened at the time, or even to find the will to do it. Scientists complicit with the regime had been skilled at covering their tracks, evidently. But maybe it was also easier for their colleagues to pretend that fellow scientists couldn't possibly have been active participants in murder and torture. Perhaps, they imagined, they were just bystanders, caught up in the mess while trying to do get their work done.

The truth—that it is perfectly possible for prominent scientists to be racist, to murder, to abuse both people and knowledge—doesn't sit easily with the way we like to think about scientific research. We imagine that it's above politics, that it's a noble, rational, and objective endeavor, untainted by feelings or prejudice. But if science is always so innocent, how is it that members of such a large and prestigious scientific organization could have sold themselves to a murderous political regime as recently as the middle of the twentieth century?

The answer is simple: Science is always shaped by the time and the place it is carried out in. And ultimately it is at the mercy of the personal political beliefs of those carrying it out. In the case of some Nazi scientists, particular experiments may have been conducted perfectly accurately and rigorously. They may even have produced good science, if goodness is measured in data and not human life. Other times researchers just didn't care about the truth or other people's lives, choosing instead to give the illusion of intellectual weight to a morally bankrupt ideology because it suited them.

Now, decades later, the horrors of World War II still have a warping effect on how we think about race science. Many of us choose to remember Nazi scientists like Otmar von Verschuer as some kind of uniquely evil exception, nothing like scientists who found themselves on the winning side of the war. The Holocaust and the twisted scientific rationale behind it are thought to belong to that time and place alone, to be purely the work of "the bad guys." But there was one question that went unanswered after the investigations into the blood-stained history of the Max Planck Society: Were scientists in the rest of the world so blameless?

To define what happened during the war as aberrant—as something that could only have been done by the worst people under the worst circumstances—ignores the bigger truth. This was never a simple story of good versus evil. The well of scientific ideas from which Hitler and others in his regime drew their plans for "racial hygiene," leading ultimately to genocide, didn't originate in Germany alone. They had been steadily supplied by race scientists for more than a century from all over the world, supported by well-respected intellectuals, aristocrats, political leaders, and women and men of wealth.

Among the most influential of them all, as far as the Nazi regime was concerned, was a pair of statisticians working not in Germany, but in the famous old literary quarter of London, in Bloomsbury, at 50 Gower Street.

| | | | | | | | | |

"You have biologists who say there is no such thing as race, we need to get over it, forget it," Subhadra Das tells me in an angry whisper. "But then, if there is no such thing, why did you just say 'race'? Where did that idea come from?"

Das is a curator of the University College London Medical and Science Collections. She moonlights occasionally as a stand-up comedian, her dark wit betraying a fury fed by the things she's learned from her research. We're in the heart of Bloomsbury, recognizable by its peaceful garden squares and smart Georgian townhouses. Once a meeting point for artists and writers, including Virginia Woolf, it is still home to a large slice of London's universities and colleges. Outside, Gower Street is jam-packed with students getting to lectures, but where Das and I are is library quiet. We're seated at a small table inside the Petrie Museum, named for Sir Flinders Petrie, an Egyptologist who, before he died in 1942, used to collect heads from around the world to shore up his ideas of racial superiority and inferiority.

"Scientists are socialized human beings who live within society, and their ideas are social constructions," she says. She wants me to hear this, setting the scene before she begins unfolding the packets of objects in front of us, which she has pulled from the archive. Among the first is a black-and-white photograph of a well-dressed older man, his bushy eyebrows resting in a canopy over his eyes, long white sideburns trailing down to his collar. Underneath is his autograph: it is the biologist Francis Galton, a young cousin of Charles Darwin who was born in 1822. Galton, she tells me, is the father of eugenics. He coined the term in 1883, from the Greek prefix *eu-* for "well" or "good," to describe the idea of using social control to improve the health and intelligence of future generations.

Galton considered himself an expert on human difference, on the finer qualities that make a person better or worse. If not quite the genius that Darwin was, he certainly aspired to be. "I find that talent is transmitted by inheritance in a very remarkable degree," he had once written in an essay titled "Hereditary Character and Talent." His logic, drawing on his cousin's theories of natural selection and the survival of the fittest, was that a race of people could be improved if the most intelligent were encouraged to reproduce and the stupidest were not—the same way you might breed a fatter cow or a redder apple. Some saw it as a way of artificially speeding up human evolution, driving the race closer to mental and physical perfection.

As an example, he drew on the fact that brilliant writers were often related to other brilliant writers. He noted that of 605 notable men who lived between 1453 and 1853, one in six were related to another member of the same group. The ingredients for greatness must be heritable, he reasoned, choosing to overlook that their being notable might be a product of wealth and connections. "If a twentieth part of the cost and pains were spent in measures for the improvement of the human race that is spent on the improvement of the breed of horses and cattle, what a galaxy of genius might we not create!" Galton dreamed of a "utopia" of highly bred superpeople, and he made creating such a perfect society his lifelong mission.

The first challenge would be to measure people's abilities, to build up a bank of data about who exactly were the most intelligent and who the least. In 1904 he convinced the University of London to set up the world's first Eugenics Record Office at 50 Gower Street, dedicated to measuring human differences, in the hope of understanding what kind of people Britain might want more of. University College, London, jumped at the chance, replying to his request within a week. After a short time it became known as the Galton Laboratory for National Eugenics.

The word "eugenics" is no longer used around here. Long after Galton's death, his laboratory was renamed the Department of Genetics, Evolution and Environment; it is now housed in the Darwin Building. And this is where Subhadra Das steps in. Among the vast collection of objects she is responsible for at the university is Galton's personal archive, which contains his personal photographs, equipment, and papers—documents that track the genesis and development of eugenics. Das also looks after objects that belonged to Galton's close collaborator, the mathematician Karl Pearson, who became the first professor of national eugenics in 1911, after Galton died. "Pearson's greatest contribution, the thing that people remember him for, is founding the discipline of statistics. A lot of work on that was done with Galton. "Galton, if you're going to bring his science down to anything in particular, is a statistician," Das tells me.

But before he settled down into science, Galton had been an explorer. He was lavishly funded by the estate of his wealthy father, who had made a fortune from the slave trade and later, when the slave trade was prohibited, from gun manufacturing and banking. An expedition to Namibia in 1850, then known as Damaraland, earned Galton a medal from the Royal Geographical Society. Always proud of his appearance (a hand mirror and sewing kit are among his personal possessions in the collection), he donned a white safari suit, becoming one of the first to cultivate what is now the classic image of the white European in Africa. "If I say to you 'African explorer,' the picture that pops into your head? That's him," Das tells me.

What was unusual about Galton was that travel failed to broaden his mind. "A lot of white Europeans, on going to Africa and living cheek by jowl with African people, tended to change their opinions," she says. "Galton didn't. If anything, his racist assumptions were made stronger by his time in Africa." As Galton told the Royal Society on his return, "I saw enough of savage races to give me material to think about all the rest of my life."

In London, racism combined in his scientific research with a passion for data. Galton was obsessed with measuring things, once using a sextant to size up an African woman's proportions from a distance. Another time he came up with the mathematical formula for the perfect cup of tea. Through eugenics he saw a way of using what he thought he knew about human difference, shored up by Darwin's theories of natural selection, to systematically improve the quality of the British race. "Darwin said that humans are animals like any other animal. Galton said, well, if that's the case then we can breed them better," Das explains. "What he was concerned about was what he saw as the degeneration of the British race and how that could be prevented and improved.

"You have to call Galton a racist because the work that he did is fundamental to the story of scientific racism. So not only is he a racist, he is part of the way we invented racism, and the way that we think about it."

| | | | | | | | | |

Eugenics is a cold, calculated way of thinking about human life, reducing human beings to nothing but parts of the whole, either dragging down their race or pulling it up. Yet somehow it seemed to make sense at the time, with a logical appeal that stretched across the political spectrum. We associate it today with the fascists who perpetrated the Holocaust, but before World War II, many on the left saw it as socially progressive. Galton himself was certainly not considered a crank. He was a fellow of the Royal Society, and an anthropometric laboratory he set up in 1884 to catalogue people's measurements enjoyed support from the British Medical Association. Eugenics belonged firmly to establishment science, and among intellectuals, it wasn't just mainstream, it was almost fashionable.

The fly in the ointment was how to make eugenics work. Galton's observation was that the poor seemed to be outbreeding the rich, and the poor were poor for the simple reason that they were congenitally unfit. Responsible action needed to be taken to address the problem and ensure genetic progress. On the one hand, the rich needed to step up their baby-making game. On the other, society's dregs, particularly those described as mentally feeble or physically weak and criminal types needed to be convinced to have fewer children. Managing reproduction was the linchpin of eugenics, even attracting a fan in the English women's rights activist and birth control pioneer Marie Stopes. To support her first clinic, Stopes founded the Society for Constructive Birth Control and Racial Progress.

The philosopher Bertrand Russell suggested that the state might improve the health of the population by fining the "wrong" type of people for giving birth. Eugenics was more than a theory—it was a plan in search of policymakers. Winston Churchill, then first lord of the Admiralty, was welcomed as vice president at the first International Eugenics Congress, held at the University of London in 1912. Other vice presidents included the lord mayor of London and the lord chief justice. Delegates came from all over Europe, Australia, and the United States, including from Harvard and Johns Hopkins Universities.

Yet despite all the support eugenics attracted from politicians and intellectuals, the field never managed to gain a firm toehold in Britain. It fell short of being implemented by the government.

The same wasn't true in the United States. The state of Indiana passed the world's first involuntary sterilization law in 1907, informed by eugenicists who argued that criminality, mental problems, and poverty were hereditary. More than thirty other states soon followed, with enthusiastic public backing. In 1910 a Eugenics Record Office was established at Cold Spring Harbor on Long Island, with support from the oil industry magnate John D. Rockefeller and, later, funding from the Carnegie Institution of Washington. Its board of scientific directors included the inventor of the telephone, Alexander Graham Bell, and the economist Irving Fisher. The hardware behind at least one of the United States' most ambitious eugenics projects came from none other than IBM, the same company that went on to supply the Nazi regime in Germany with the technology it needed to transport millions of victims around concentration camps.

By 1914 the word "eugenics" was being used with such abandon that it had almost became synonymous with being healthy, Roswell H. Johnson, a professor of eugenics, complained in the *American Journal of Sociology*. "A school for sex education is called a school of eugenics. Even a milk and ice station has been similarly designated," he grumbled.

As all this went on, a few couldn't help but notice the holes in this grand idea. It was a slippery slope. Henry Maudsley, a psychiatrist, pointed out that privilege and upbringing could surely more accurately explain why some people were successful and other weren't. He noted that many remarkable people, including William Shakespeare, had unremarkable relatives. Another vocal critic of eugenics was the biologist Alfred Russel Wallace, who had come from humble beginnings to become an important and well-loved researcher and was credited with coming up with the theory of evolution by natural selection at the same time as Darwin. "The world does not want the eugenicist to set it straight," he warned. "Give the people good conditions, improve their environment, and all will tend towards the highest type. "Eugenics is simply the meddlesome interference of an arrogant, scientific priestcraft."

| | | | | | | | |

In its early days, particularly for its mainstream supporters, eugenics focused on improving the racial stock by weeding out those at the margins: the feeble-minded, the insane, and the disabled. But as time wore on the umbrella inevitably expanded. Karl Pearson, who succeeded Galton as the main force behind eugenics after the latter's death in 1911 and shared his views on race, believed that since other races were inferior to his own, intermixing was also dangerous to the health of the population. By this logic the very existence of these other races represented something of a threat. "Pearson's argument is that if you have uncontrolled immigration the welfare of British people is at stake," Das tells me.

At this point, she pulls out another object from the archive. It's a narrow tin box resembling a cigarette case but twice as long. It was brought to London by Pearson, but was designed by Eugen Fischer, a German scientist who had been director of the Kaiser Wilhelm Institute of Anthropology, Human

Heredity, and Eugenics. The box still bears Fischer's name. Inside are thirty locks of artificial hair in a neat row, ranging in color from bright red hair at one end to blond in the middle (numbers 19 and 20), then light brunette, and kinky black hair (number 30) far at the other end. At first glance it looks innocuous, like a color chart you might find at the hairdresser's. But the disturbing story behind it is betrayed by the order in which the hair samples are placed. The most desirable colors and textures have been placed in the middle and the least acceptable at the margins. This simple little gauge tells a story of pure horror.

"Fischer used this device in Namibia in 1908 to establish the relative whiteness of mixed-race people," she says. In what is now remembered as the first genocide of the twentieth century, in the four years of the run-up to 1908, Germany killed tens of thousands of Namibians when they rebelled against colonial rule. According to some estimates, up to three thousand skulls belonging to members of the Herero ethnic group were sent back to Berlin to be studied by race scientists. "Namibia was the first place where the Germans built a concentration camp. Depending on where your hair fell on the scale was the difference between life and death." Similar methods would be used again a few decades later. Fischer's work informed the Nuremberg Laws of 1935, which outlawed intermarriage between Jews and blacks and other Germans. He became a member of the Nazi Party in 1940.

Das takes out another box that had belonged to Pearson, this one containing rows of glass eyes in different colors, framed in aluminum eyelids so eerily real that I fear one of them might blink. They are prosthetics of the kind that would have been fitted in patients who had eyes missing. In the context of eugenics, though, they served another purpose. "This object, I have seen its twin brother on display in an exhibition about race hygiene in Germany at the Berlin Museum of Medical History at the Charité. This device was appropriated by Nazi scientists and, again, used to judge or measure race, particularly in Jewish people," she explains. "You'll find photographs of Nazi scientists measuring people's heads, measuring people's noses, matching their eye color."

The eye and hair color charts reveal just how slippery the dogged mantras of rationality and objectivity can be when it comes to studying human difference. "Any scientist who claims that they are not politicized, or that they are asking questions out of pure curiosity, they are lying to themselves," she continues. "The structure in itself is fundamentally, structurally racist, because it has always been taken it at its face. Never going back and taking apart those underpinnings." What does it matter if one person has black hair and brown eyes, and another has blond hair and blue eyes? Why not compare heights or weights or some other variable? These particular features matter only because they have political meaning attached to them.

In the first decades of the twentieth century, all over the world, eugenics began to be conflated with old nineteenth-century ideas about race. In Japan, during the Meiji period, the thinker and politician Katō Hiroyuki used Darwinism to make the point that there was a struggle for survival between different nations. In China in 1905, the revolutionary Wang Jingwei argued that a state made up of a single race was stronger than one comprising multiple races. Other Chinese politicians advocated sterilization as a means of human selection, and racial intermarriage to produce children with whiter skins. Historian Yuehtsen Juliette Chung has noted that during this time, "China seemed to accept passively the notion of race as the West understood it."

In India, too, European notions of racial superiority were easily absorbed by some, partly because they mirrored the country's existing caste system—itself a kind of racial hierarchy—but also because Germany's Aryan myth placed the noble race as once living in their region. The ideological quest for the true "Aryans" remains alive in India, and Adolf Hitler's *Mein Kampf* is a bestseller in Indian bookshops. Each nation used the idea of race in its own ways, marrying it with science if it could be of use. Eugenics, then, became just another tool in what were longstanding power dynamics.

In the United States, arguably the most racially charged place in the world at the time, evolutionary theory and eugenics came along at just the moment that intellectual racists could deploy them to full effect. When the Eugenics Record Office was opened at Cold Spring Harbor, New York, a news item in the journal *Science* announced that one of the purposes of the new office would be "the study of miscegenation in the United States," the mixing and intermarriage of different racial groups. Immigration into the US from countries considered to be undesirable had already been curbed by the 1882 Chinese Exclusion Act, the country's first major law restricting immigrants. Twelve years later, three Harvard College graduates, lobbying in favor of a literacy requirement for those who wanted to come to the United States, formed the Immigration Restriction League. The group's secretary, Prescott Farnsworth Hall, used Darwin's ideas on natural selection to argue against "undesirable" immigrants who weren't "kindred in habits, institutions and traditions to the original colonists." In a lengthy racist tract in *Annals of the American Academy of Political and Social Science* in 1904, he wrote, "The doctrine is that the fittest survive; fittest for what? The fittest *to survive in the particular environment in which the organisms are placed*" (Hall's emphasis).

By 1907 riots in Bellingham, Washington, saw hundreds of white men, themselves recent arrivals from Europe, attack Indian immigrants who lived in the city, blaming their "filthy and immodest habits." Reportedly, seven hundred Indians had to flee. The local *Bellingham Herald* editorialized, "The Hindu is not a good citizen. It would require centuries to assimilate him, and this country need not take the trouble."

It was against this backdrop that a new ideologue emerged. In 1916 a wealthy American law graduate named Madison Grant published a book that took eugenics to another level. Grant was known as a conservationist: as one of the founders of Bronx Zoo in New York, he had lobbied to put the Congolese Ota Benga on display among the apes there in 1906. Grant wasn't a scientist, but he recognized the power of the language of science. In *The Passing of the Great Race: or The Racial Basis of European History*, he continued the legacy of Count Arthur de Gobineau from the previous century, pushing forward the myth of Aryanism. Grant proposed that a blond, blue-eyed Nordic "master race" represented the Aryans' true descendants.

His racial hierarchy was geographically specific. Everyone who wasn't northern European was consigned to an inferior status, including Italians and Greeks, who at that time were considered an undesirable immigrant group in the United States. Grant warned against racial intermixing in the belief that this would damage white racial purity even further. As casually as a biologist writing about plant hybrids, he wrote that a cross between any member of a European race and a Jew is a Jew.

In Grant, wealth and racism formed a toxic combination. As someone descended from some of the first European colonists to settle in America, he of course counted himself among the descendants of Aryans, a noble race under threat. Openly in favor of both slavery and segregation, he made every possible effort to cut immigration to the United States from anywhere but northern Europe. And he had powerful supporters, including Theodore Roosevelt, soon to be president. Grant became vice-president of the Immigration Restriction League in 1909, and Roosevelt was also a member. In 1921 Grant was the treasurer at the Second International Eugenics Conference in New York.

It took only the slightest interrogation for Grant's historical and scientific evidence to be exposed as dodgy and self-serving. One reviewer of Grant's book raised an eyebrow at his claim that the Italian artists Dante,

Raphael, Titian, Michelangelo, and Leonardo da Vinci were of the Nordic type, and that—stretching the geographical parameters even further—so was Jesus. But the views of experts didn't matter to many of Grant's readers. His fake assertions were enough for those looking for what they wanted, some seemingly intellectual support in their opposition to immigration.

Two parallel ideologies had by now firmly intertwined in the minds of racists. One, the decades-old concept of the existence of a superior race. The second, informed by eugenics, that unless checked, inferior races would outbreed superior ones. Human variation had gone from being, before the eighteenth century, a jelly-like set of loose generalizations to a hard matter of progress and struggle. Grant's work was referenced by the Ku Klux Klan. It also became one of the inspirations behind the Immigration Act of 1924, which set quotas according to nationality aimed at decreasing immigration from southern and eastern Europe, including Italy, Greece, and Poland, as well as effectively barring anyone from Asia.

And his work earned one lifelong fan in Germany. In a fawning letter to Grant about *The Passing of the Great Race*, Adolf Hitler wrote, "The book is my bible."

| | | | | | | | |

It was all so long ago—we imagine that it's well and truly over now. We think of the horrors of the Holocaust and earlier genocides, of slavery and colonialism, of the many millions who were killed, of the twisted logic behind these actions, as belonging to another time. We imagine that the end of World War II spelled an abrupt end for race science. Eugenics is a dirty word. We're enlightened now. We're wiser.

But the story doesn't end quite so quickly. Although they may have tempered their politics, race scientists didn't simply disappear after the war. Those who had built their work around eugenics and studying human difference, who staked their careers on these studies, just found new avenues.

Take Otmar von Verschuer, who had plundered the tiny bodies of Auschwitz victims for his twin studies during the Holocaust: after being banned from teaching temporarily, in 1951 he became professor of human genetics at the University of Münster. Many scientists similarly changed tack, gently maneuvering themselves out of eugenics into allied fields that studied human difference in less controversial and more rigorous ways, such as genetics. Many stopped using the word "race" altogether. Science learned at least one lesson: if scientists wanted to study human variation, they had to try to stay away from politics.

But the shift didn't happen abruptly. The Eugenics Record Office on Gower Street in London survived all the way through the war. There is still a Galton Professor of Genetics at University College London, funded by the money Francis Galton left to the institution. What was once the Eugenics Society became the Galton Institute in 1989. In 2016 the institute established the Artemis Trust, which according to its promotional leaflet, handed to me at a conference, distributes grants of up to £15,000, partly with the aim of assisting in the provision of fertility control, particularly to those from "poorer communities."

Subhadra Das tells me that a woman came to see her recently whose mother had worked in the Galton laboratory in the 1950s. Her job had been to study redheadedness in Wales. It took until the 1960s for the word "eugenics" to stop being heard in these corridors. What actually helped kill it in the end wasn't just the war but also the fact that new research showed it couldn't actually work. The genetics around inheritance, once it was better understood, didn't support the idea that humans could breed themselves to perfection, whatever perfection meant. The way we inherit traits from our parents turns out to be more complicated than Galton imagined. There is actually no guarantee that two beautiful and brilliant parents will produce brilliant and beautiful kids. Genetics is bit more of a crapshoot.

Yet eugenics policies introduced to other parts of the world took decades to be shut down. Only in 1974 did Indiana repeal legislation that had made it legal to sterilize those it considered undesirable. Investigations by the reporter Corey Johnson in 2013 uncovered that doctors working for the California Department of Corrections and Rehabilitation had continued the practice, sterilizing as many as 150 women inmates between 2006 and 2010, possibly by coercing them into having the procedures. In Japan, a Eugenic Protection Law that was introduced in 1948 to sterilize those with mental illness and physical disabilities and prevent the birth of "inferior" offspring was stopped only in 1996. Victims of the legislation are still pushing for justice.

The process of self-examination, of experiencing regret and showing remorse—the kind attempted by the Max Planck Society in 2001—is slow. And it has been particularly slow in the places that found themselves on the winning side of World War II. In the decades after the war, scientists in Britain and the United States airbrushed away their pivotal role in race science and eugenics. Scientists quietly moved into other fields, silently renamed their university departments, consigning to the past that dark chapter. History was rewritten by the victors.

According to Gavin Schaffer, a professor of British history at the University of Birmingham and the author of *Racial Science and British Society, 1930–62*, "It was much easier to point the finger at the horrible Nazis, and the same went for the scientists. This absence of introspection was rooted in the ability to point fingers at other people for being responsible for the perversion of science."

The postwar narrative of good triumphing over evil glossed over the messier truth: that in fact everyone who pointed a finger at others should have pointed a finger at themselves. Without ever really looking back to the past and asking how and where the idea of race had been constructed in the first place, why it had been relentlessly abused—without questioning the motives of scientists such as Francis Galton, Karl Pearson, and countless others—in this glaring "absence of introspection," old ideas of race could never completely disappear. Even long after the war, scientific fascination with human variation remained tainted by a lingering belief that there might be something deeper about racial difference, that perhaps some races really are better than others.

Yes, some good science emerged from the ashes. Biology did attempt to reform itself, to cast away the mistakes of the past and do a more precise and accurate job of understanding human variation. But at the same time, while the world around them changed, a few of the hardened old-school race scientists could still be found knocking about. "Racist science continues; it just becomes more marginal," Schaffer tells me. "But there's no doubt that it does continue."

CHAPTER 4

# Inside the Fold

*After the war, intellectual racists forged new networks*

The Second World War marked an unlucky turning point in the life of Reginald Ruggles Gates.

Born in 1882, Gates was one of those well-to-do, gentlemanly race scientists who were so common in the nineteenth century. He was a colonial type who believed that other races were different human species and a eugenicist who supported segregation in the United States. He would certainly be considered a racist by modern standards, but at the time his views weren't uncommon. They didn't get in the way of his career or his standing in society. He was successful and well respected.

To get a sense of who Gates was, I've come to the Maughan Library at King's College London, where his archive was moved after he died in 1962. It is off Chancery Lane, housed in a vast nineteenth-century Gothic Revival building that was once Britain's Public Record Office. I leaf through his personal papers, slowly building a mental portrait. Sepia photographs show him to be smartly dressed, sporting a neatly clipped moustache. Gates had grown up in a wealthy family with thousands of acres of land across Nova Scotia, Canada, before he moved to Britain, where he was briefly married to Marie Stopes, a fellow member of the Eugenics Society. He enjoyed a career as a plant geneticist, becoming professor of botany at King's College in 1921, and later a fellow of the Royal Society. He seems to have had a passion for travel, too, for understanding human difference across the world. His collection of scientific papers spans work from almost every continent.

Before the war he was clearly riding high. After the war, though, it all changed.

Gates found, to his confusion and disappointment, that he was now being left out in the cold by an establishment that had once welcomed him. His papers were rejected by scientific journals more than they had ever been. The reason was simple: deeply shaken by the genocidal use of eugenics by the Nazis in Germany, the world was turning its back on any research that resembled their theories of racial hygiene. The enthusiasm for studying race—once almost fashionable in scientific circles—was on its way out. Researchers who, like Gates, weren't wise enough to get with the new program, who chose instead to cling to their unpalatable politics, found themselves flung from the warm center of academic life to its chillier margins.

Yet he couldn't figure out why. "What interests me about it is his incredulity," I'm told by Gavin Schaffer. "He seemed genuinely surprised." In a sense, Gates was a man caught out by time. While Francis Galton and Karl Pearson had the good fortune to die before they could witness race science reach its most brutal peak, others survived long enough to see the political mood change—and then suffer the consequences.

At every opportunity, Gates refused to budge from his belief in racial superiority and inferiority. Wherever he found himself professionally hindered, he imagined himself to be the victim of a Jewish plot to derail his work. Schaffer recounts one especially bad experience, in 1948, when Gates was working briefly at Howard University, the historically black college in Washington, DC. "A petition was got up to remove him because of allegations that he was a racist—which he *was*. But he was stunned by that," Schaffer says. "He articulated his understanding of that as a manifestation of an international Jewish conspiracy, as opposed to just understanding that, in a historically black university, the kind of work that he did and the kind of things that he said were always going to be challenged." Even when he agreed to leave Howard, Gates grumbled in private that only a few "ignorant Negroes" were fit to be in a university at all.

In later life he turned to travel, his wide-ranging interest in human differences taking him all over the world. He visited Cuba and Mexico to study "mixed-race" people and Japan and Australia to observe indigenous communities, and made a number of trips to India, a country that became a particular source of fascination. Browsing his personal collection of scientific papers in the library, I'm startled to discover there's even one on the blood groups of the Sainis in parts of Punjab—a study that may well have included relatives from my father's branch of the family.

What Gates could never accept was that the world was moving on, leaving those like him behind.

| | | | | | | | |

When it came to how the world thought about race, a wider political shift was under way. It was most clearly signposted in 1949 when more than a hundred scientists, anthropologists, diplomats, and international policy makers met in Paris under the umbrella of the United Nations Educational, Scientific and Cultural Organization, UNESCO, to redefine race. A British-born American writer and anthropologist, Ashley Montagu, led the charge against scientific racism and its horrific legacy, taking his cue from a wave of social scientists who had already long argued that history, culture, and environment were really behind what people thought of as racial difference.

"The word race is itself racist," Montagu wrote in his influential 1942 book *Man's Most Dangerous Myth: The Fallacy of Race*. Both intellectually and culturally ahead of the curve, he explained in an article in *American Anthropologist*, "What a 'race' is no one exactly seems to know, but everyone is most anxious to tell. . . . The common definition . . . is based upon an arbitrary and superficial selection of external characters." As anthropologists and geneticists were learning, individual variation within population groups, overlapping with other population groups, turned out to be so large that the boundaries of race made less and less sense. This was one reason why nobody had ever been able to agree on exactly how many races there were. Three, or four, or five, or several, there was never a consensus. The concept of race was as slippery as jelly, defying any effort to pin it down. In the end, academics had to concede that it probably wasn't an accurate or reliable way to think about human variation.

Montagu emphasized the likelihood that humans were genetically pretty much identical, and that in any case, our ancestral roots were certainly the same. Other anthropologists who had studied human diversity had already suggested that differences between humans were not only marginal but also formed a continuum, each so-called race blurring into the next. What really made people and nations seem different was culture and language, neither of which is hereditary.

It was on the back of work like this that UNESCO, in July 1950, released its first statement on race, stressing unity between humans in a concerted effort to eradicate what it saw as the outcome of a "fundamentally anti-rational system of thought." It was meant to be the last word on the subject, to flush away racism once and for all. "Scientists have reached general agreement in recognizing that mankind is one: that all men belong to the same species, *Homo sapiens*."

The next few decades would be crucial to dismantling the idea that race was real and to proving Montagu right. In 1972 a landmark paper exploring the true breadth of human biological diversity appeared in the annual edition of *Evolutionary Biology*, written by a geneticist, Richard Lewontin, who later became a professor at Harvard University. Dividing the planet up into seven human groups, based roughly on old-fashioned racial categories, Lewontin investigated just how much genetic diversity there was within these populations compared with the genetic diversity between them. What he found was that there was far more variation among people of the same "race" than between supposedly separate races; he concluded that around 85 percent of all the genetic diversity we see is located within local populations—93 percent if you widen the net to continental populations. In total, around 90 percent of the variation lies roughly within the old racial categories, not between them. There has been at least one critique of Lewontin's statistical method since then, but geneticists today overwhelmingly agree that although they may be able to use genomic data to roughly categorize people by the continent their ancestors came from (something we can often do equally well by sight), by far the biggest chunk of human genetic difference is indeed found within populations.

Lewontin's findings have been reinforced over time. An influential 2002 study published in *Science* by a team of scientists led by geneticist Noah Rosenberg, then at the University of Southern California, took genetic data from just over a thousand people around the world and showed that indeed as much as 95 percent of variation is within the major population groups. Statistically this means that although I look nothing like the white British woman who lives next door to me in my apartment building, it's perfectly possible for me to have more in common genetically with her than with my Indian-born neighbor who lives downstairs. Being of the same "race" doesn't necessarily mean we are genetically more similar.

In the long run, then, Ashley Montagu's position on race has been vindicated.

Mark Jobling, a respected professor of genetics at the University of Leicester, tells me that if there were a global catastrophe and all life were wiped out save just, say, Peruvians, 85 percent of human genetic diversity would be safely retained. "That just reflects the fact that we are a young species," explains Jobling. *H. sapiens* is relatively new, and being so new, we're still closely related to one another.

The greatest genetic diversity within *Homo sapiens* is found in Africa, because this continent contains the oldest human communities. When some of our ancestors began to migrate into the rest of the world sometime between 50,000 and 100,000 years ago, the groups that moved were genetically less diverse than the ones left behind for the simple reason that they were made up of fewer people.

The human variation we see across different regions today is partly the result of this "founder effect." Of course, groups of people have average physical differences, as a result of their biological and environmental histories. It has been estimated that ten thousand generations separate every single one of us from the original little band of people in what is now Africa, but we look different because of the characteristics we happened to take with us we migrated. As these small migrant populations spread, bred, and adapted to their local environments, they began to look ever more different from the relatives they left behind generations earlier and more like each other. And as small members of these groups, again, left for new territory, they would become slightly genetically different again because of a serial founder effect.

All of this didn't happen in big clumps or clusters, but rather like more of a mesh, as people mated with those they encountered on the way, sometimes traveling further away and sometimes moving back. If everyone in the world had their genomes sequenced, says Jobling, you wouldn't find hard borders between them, but gradients, with each small community blending into the next, the way hills blend into valleys. The racial categories we are used to seeing on census forms don't map onto the true picture of human variation.

The aim of the original 1950 UNESCO statement wasn't just to set out the science in a clear way, but also to change the culture, to make people think differently about this idea that they had lived with for so long, that had done incalculable damage to millions of lives. The statement emphasized that what we see as race is likely to be only a superficial variation on one theme. Most of the visible difference is cultural. The UNESCO statement poured cold water on entrenched racial stereotypes. Furthermore, it made clear that there was no proof that different groups of people differed in their innate mental characteristics, including intelligence and temperament.

It marked a crucial moment in history, a bold universal attempt to reverse the deep-seated damage that had been done by racism—and perpetuated by science—for at least two centuries. And to some extent, it worked. Whether we realized it or not, all of us thought about race differently after that. Racism was no longer fashionable. Scientists and anthropologists by and large got behind UNESCO, and their work in the coming decades would largely reflect that.

But this wasn't the end of it.

| | | | | | | | | |

Despite the changing public mood around race, some researchers just couldn't bring themselves to ditch a body of work they had been cultivating for decades. Many didn't agree with UNESCO's claim that biology supported the idea of a universal brotherhood. A few couldn't accept that there were no mental differences between racial groups. And they weren't all racists. Some of them were respectable, eminent scientists at universities, including Oxford and Cambridge, who simply wanted the statement to be revised with more scientific precision and qualification.

But one of the most passionate voices of all belonged to Reginald Ruggles Gates.

"What Gates called for, time and again, was the objective continuing study of race . . . because he thought that his position was grounded in *true* science," explains Schaffer. He and others felt that UNESCO was stepping outside the bounds of what biology could actually claim, that it was ignoring facts in favor of liberal, antiracist politics. "The biologists who countered it, what they wanted was the continuation of their own expertise, which they had asserted over twenty, thirty years. They agreed that the Nazi state was completely wrong in the way it had used race, and that other political actors had been completely wrong, but they felt that the study of race would profit from further work. I think, to them in that period, they felt they wanted work on race to continue, and they felt that other people wanted it to stop."

The pressure worked, at least in part. In 1951 UNESCO got a team of experts together to publish a new statement whose language was tempered to account for the lack of consensus around the biological facts. The changes were subtle, but revealing. For instance, instead of saying that scientists had "reached general agreement" that we were one human species, the revised statement was gently altered to say that scientists were "generally agreed."

In short, it had to make clear that not every expert could accept even the most basic fact that we all belonged to the same species.

Despite concessions like these, Gates failed to keep race science alive in the same way as before. By now it had been all but lifted out of the laboratory. The academic study of race no longer had a place within the realms of biology. Whether all biologists liked it or not, by the second half of the twentieth century, race belonged to the social sciences, to the study of culture and history. It was understood to be a social and political construction, not a concept borne out by biology. Old-fashioned race researchers and eugenicists had to move on or be sidelined.

At that point, Schaffer explains, "The biologists just go into themselves a bit. They go back to their work, they go back to their labs." They move into newer fields, such as genetics, evolutionary biology, and psychology. They also begin to look for difference at the molecular level rather than at the surface. "As long as you weren't hell-bent on the kind of politics that were going to call your position into threat, yes, why not." The older, somewhat cruder, and more controversial ways of studying human difference, using anatomy and twin studies, were now treated with suspicion. By the 1950s the word "race" was so unfashionable in scientific circles that it was barely used anymore.

Veronika Lipphardt, a historian at University College Freiburg, in Germany, has noted that the 1950s saw new institutes dedicated to the study of human variation open around the world. There was one in Bombay, another at Columbia University in New York, and one at the Federal University of Paraná in Brazil. A politically correct scientific terminology emerged. Researchers began referring to groups as "populations," and occasionally as "ethnic groups." But the departure from the old race science wasn't quite as complete as it might have been. Although the parameters of research had changed, the racial categories were still alive in people's minds. They were still active in people's everyday lives, playing out in the racism of the real world. For scientists to suddenly stop thinking about humans in racial terms was impossible so long as everyone out there still thought about themselves and others that way. So they couldn't help but look for racial difference, to subconsciously force this way of thinking into their work.

One example is blood type. When genetics started to become the preferred way to talk about human variation, hard hereditary variables such as blood type came under the spotlight. Categorizing by blood sounded more mathematical, less wishy-washy, than talking about skin color or hair texture. And in the process, blood became an obsession. It was already well known that the proportions of people with different blood types varied from population to population, because of a phenomenon known as genetic drift. In prehistory, as small founding communities of people migrated across the world, they took their own narrow subset of blood types with them. This is equivalent to your cousin, say, leaving home to set up a colony of her own. She may be closely related to you, but have a different blood type. As these communities got bigger, their particular blood types became the common ones. For example my own, B+, is shared by around a third of people in India, where my family are from, and less than a tenth of people in the United Kingdom, where I live. By studying which population groups have which blood types, researchers found they could open a window into how closely or distantly related these groups might be to each other.

In the postwar period the distribution of blood types became a hot topic in anthropology journals. In the 1960s the World Health Organization launched its own effort to document groups of people around the world, collecting data on skin pigmentation and hair form, but also on blood type, color blindness, and other genetic markers. When the blood of the Sainis of Punjab was collected and tested in 1961 by a pair of anthropologists at the University of Delhi, before the results were passed to Reginald Ruggles Gates in England, it was part of these bigger efforts. Similar tests were carried out in other Indian groups and castes. Ultimately thousands of people from different communities were gauged in the same way, and the same occurred all over the world. At least some of these scientists were searching for proof that race was real, that evidence for racial differences could be found at the molecular level.

Gates was one of these scientists. He just couldn't let go of his belief in the old, hard, biologically rooted racial gaps. And he would never change. His final piece of work, entitled *The Emergence of Racial Genetics*, published posthumously in 1963, attempted to place the new genetics in the old framework of race. Part of the reason he pressed on with his commitment to races as meaningful categories, argues Schaffer, is that he sincerely believed that his own research was objective and that those challenging him were the ones driven by ideology. He saw himself as the bearer of truth, held back by an antiscience political agenda that was mistakenly trying to impose racial equality on the world. All the while he was receiving funding from segregationists in the United States.

Schaffer reminds me that it's important to understand the psychology behind Gates's conviction. When Gates complained that race research was being politicized, he was right. After all, following World War II and the bald brutality of the Holocaust, it would have been bizarre for any discussion of race not to be affected by politics. But what Gates failed to accept was that he was similarly affected by his own politics. "People like him must also position themselves within that model. People who defend race historically have also done so for political reasons," he explains. "The science never becomes separated from political discourse."

Gates wasn't a pseudoscientist, even if he was a bit of a crank. But his failure to be professionally recognized wasn't just a result of his beliefs. Editors of some of the scientific journals that rejected his work warned him that his methods were getting sloppy: he was relying on subjective interpretation rather than rigorous, intensive study. This slapdash approach may have been acceptable in the previous century, but it didn't cut it in the world of modern science. By the end, Gates had few supporters left in the scientific community, as a result of both his abhorrent views and his weak research. When his death was announced in 1962 at a meeting of American anthropologists, reportedly there were cheers.

That said, history doesn't move in a straight line. Ideas, even the worst ones, can go out fashion in one century and come back in another. Those who imagined that the end of World War II marked the abrupt death of race science were sadly mistaken.

| | | | | | | | |

In the final years before his death, when few scientific journals would touch his work, Gates decided to take matters into his own hands. If they wouldn't publish him, he would publish himself.

He and a handful of like-minded researchers, some on the very darkest margins of science—including the former Nazi scientist Otmar von Verschuer and a British eugenicist, Roger Pearson, in 1960, established a journal of their own. Their aims were simple: to challenge what they saw as a politically correct, left-wing conspiracy around race and bring back some scientific objectivity. (Von Verschuer died in a car accident in 1969, soon after Gates. Pearson, the last of the group still alive, aged ninety in 2018, declined to give me an interview for reasons of ill health.)

They named their brave new enterprise the *Mankind Quarterly*.

The founders of this revolutionary journal regarded themselves as "the defenders of the truth," says Schaffer; they even compared it to the Gospels.

But to anyone who read it, it would have been immediately clear that the *Mankind Quarterly* was not as impartial as it claimed to be. In the early 1960s, when it was launched, South Africa was an apartheid state, the US civil rights movement was gaining momentum, and European colonies in Asia and Africa were winning independence. Race was high on the agenda everywhere, and the moral failures of the past were slowly being redressed. For racists who didn't welcome this shifting tide, now was the moment to assert their position. And the *Mankind Quarterly* was happy to oblige. It waded deep into the politics of the time, using science—even if only in a loose way—as its weapon of choice.

Recruiting truly respectable scientists to the cause was a challenge, but not impossible. The earliest editions included articles by Henry Garrett, a former president of the American Psychological Association and the head of Columbia University's psychology department. He was then one of the most powerful and eminent voices against desegregation in the United States. Most notably, in 1954, Garrett had testified to stop the integration of black and white schools in the state of Virginia, when a trial known as *Davis v. County School Board of Prince Edward County* went up to the Supreme Court. Hundreds of students at an underfunded all-black school, with no gym or cafeteria, who were sometimes forced to study in an old school bus, fought against separate schools on the grounds that they were being disadvantaged because of their color. The judge ruled against them. Under national pressure, this case went on to be combined with five other school desegregation cases brought before the Supreme Court in 1954 in the landmark case *Brown v. Board of Education*, which finally declared that separate schools for black and white students were unconstitutional.

Writing in the *Mankind Quarterly* in 1960, Garrett wasn't rolling back from his defeat, he was doubling down. Regardless of what the law said, for him, people of different races mixing with each other spelled certain disaster. "The weak, disease-ridden population of modern Egypt offers dramatic evidence of the evil effects of a hybridization which has gone on for 5000 years. In Brazil, coastal Bahia with its negroid mixtures is primitive and backward as compared with the relatively advanced civilization of white southern Brazil," he wrote.

In another article for the journal, in 1961, Garrett laid into academics, politicians, and social reformers who didn't accept the "common-sense" judgement that "the Negro was . . . less intelligent and more indolent than the white." Like so many scientific racists before him, Garrett argued that this difference was evidence of civilizational superiority. He claimed that Africans had never produced anything of any great value. Could any African Negro, he charged, "compare with the best of the European whites: to compare, for example, with Aristotle, Cicero, Thomas Acquinas [*sic*], Galileo, Voltaire, Goethe, Shakespeare or Newton?"

The twisting of facts to suit an ideological viewpoint would become a regular feature of the *Mankind Quarterly*. An especially cold-blooded article in 1966, by one of the journal's editors, argued that Aboriginal Australians and Native Americans had been all but wiped out by European colonizers not out of greed or cruelty, but because it was a natural outcome of biology. "If the conquered are markedly inferior to the conquerors . . . they will always remain an outcaste element at the bottom of the social structure." Interracial conflict, the writer argued, was the product of natural selection. Drawing parallels with the American civil rights movement, he added that it was virtually self-evident that racial integration would never work.

Articles like these didn't go unnoticed in the scientific community. Almost as soon as the *Mankind Quarterly* appeared, disgusted anthropologists sent in letters of complaint, accusing the journal of trying to make scientific racism respectable again. A Slovene anthropologist, Božo Škerlj, who had mistakenly joined the *Mankind Quarterly*'s advisory board, only to be appalled by its "ostensibly racialist editorial policy," entered into a public spat with the editors. Škerlj was particularly insulted by Gates's accusation that his mental outlook—and presumably his objectivity—was affected by the fact that he had been imprisoned in the Dachau concentration camp during the war. Gates noted, revealingly, that he would never have considered him for the position in the first place had he known about his internment.

In *Science*, one of the world's leading journals, a reviewer called on scientists to take action "against this unwelcome, ill-founded unbiological outgrowth of racism." But it made no difference. The reason the *Mankind Quarterly* had been created at all was the lack of scientific approval for the kinds of ideas its founders wanted to publish. The editors didn't want or care about approval, they just needed a platform.

The other, deeper secret behind the *Mankind Quarterly* was that it had legs of its own. Support came indirectly from a reclusive, multimillionaire textile heir with a vested political interest in the articles the journal published. Wickliffe Draper was a diehard segregationist descended from a commanding officer in the Confederate Army on one side and the largest slaveholder in the state of Kentucky on the other. His family roots in North

America dated back to 1648, with enormous wealth and property steadily amassed over the centuries. In his 2002 book *The Funding of Scientific Racism*, William Tucker, emeritus professor of psychology at Rutgers University, details Draper's upbringing, one so privileged that a relatively weak academic record wasn't enough to prevent him from getting into Harvard. In the early twentieth century at Harvard, Tucker writes, Draper would have been exposed to those at the forefront of the country's eugenics movement. Draper was an intellectual racist looking for ways to spend his inheritance; the *Mankind Quarterly* would turn out to be the perfect vehicle for his racist views.

In March 1937 Draper incorporated the Pioneer Fund, a private foundation whose purpose was to disseminate information on human heredity and eugenics and provide race scientists who couldn't find backing anywhere else with the cash they needed to carry on. The anthropologist Robert Wald Sussman explains in his 2014 book *The Myth of Race* that "Draper wanted to recruit scientific authorities with academic credentials and scholarly records who believed in the necessity of racial purity and [believed] that integration posed a threat to civilization." In short, he was trying to build a scholarly argument to defend segregation. During the war his money was used to help distribute a Nazi propaganda film about eugenics to US schools and churches. But it was after the war that the fund really came into its own. In 1959 Draper set up what he called the International Association for the Advancement of Ethnology and Eugenics, to produce and publish documents on race. The association made it its aim to promote and distribute the *Mankind Quarterly*, to help turn it into one of the most important vehicles for race research in the world.

Tucker describes the original directors of the International Association for the Advancement of Ethnology and Eugenics as "probably the most significant coterie of fascist intellectuals in the postwar United States and perhaps in the entire history of the country."

The Pioneer Fund's priority from the beginning was to back distinguished scientists, the more well known the better, along with racist ideologues. "Grants to the former were intended to provide a façade of intellectual respectability for the latter, as well as results that could be used to justify their policies." Cash gifts were routinely made to scientists who echoed Draper's political sentiments, while thousands of copies of the *Mankind Quarterly* containing their work were sent out to a list of American political conservatives. The science and the politics operated hand in glove.

Unsurprisingly, then, according to Tucker, the journal made no concessions to political correctness. "This was going to be a publication frankly written *by* racists *for* racists," he writes. The target audience didn't appear to be the academic community at all, but racist movements searching for evidence that their prejudices might be rooted in scientific fact. "Nothing seemed too bizarre or too repugnant to receive the *Mankind Quarterly*'s stamp of approval." One of the longest articles it published was titled "The New Fanatics," which slammed American intellectuals who used their authority to support equal rights for black people. Sussman has noted that the book review section was in essence a bulletin board for publications that had anything to do with eugenics, where praise was lavished on new publications that were neo-Nazi, anti-Semitic, or antiblack.

The editors' intentions would have been clear to the journal's readers. Many of the mainstream scientists who did bother to read the new journal saw straight through it. A scholarly review of the first three editions by the late British anthropologist Geoffrey Ainsworth Harrison, a former president of the Royal Anthropological Institute, was scathing. He complained that one of the editors hadn't grasped the concepts of modern genetics, despite writing about them at length. He dismissed Henry Garrett's work, too, as full of inconsistencies. He even threw in a comment about how many typographical errors there were. Although Harrison saw some value in studying human variation, he didn't see what the *Mankind Quarterly* did as academically useful. "Few of the contributions have any merit whatsoever, and many are no more than incompetent attempts to rationalise irrational opinions. . . . It is earnestly hoped that *The Mankind Quarterly* will succumb before it can further discredit anthropology and do more damage to mankind," he concluded.

But that didn't happen. In fact, the journal kept going for many more decades, publishing scientists and not-quite-scientists at the margins of their fields, many of them personally bankrolled by Wickliffe Draper's Pioneer Fund. The fund stuck to its aims even after Draper died in 1972. In his will, Draper left $50,000 to Henry Garrett alone. Despite all the criticisms it faced when it was first published, despite the widespread expectation that it wouldn't last, the *Mankind Quarterly* never succumbed.

If you want to read it, it's still around today.

| | | | | | | | |

When I contacted the German biochemist Gerhard Meisenberg, the current editor in chief of the *Mankind Quarterly*, I didn't expect to hear back from him.

After all, this is a journal considered so inflammatory that the security filter of my home broadband provider in the United Kingdom automatically blocks it. So I'm surprised to not only hear from Meisenberg immediately, but that he seems perfectly happy to tell me whatever I want to know. He advises me that he became the editor of the *Mankind Quarterly* only within the last few years, and that his job is to "start the seemingly hopeless task to salvage this rundown journal," betraying the possibility that he sees renewed interest for the ideas they publish. At the same time, he warns that he can only communicate with me through email—not because he considers the work he does to be inflammatory, but because it's tricky to reach him by any other means right now. Since 1984 he's worked at the Ross University School of Medicine, a for-profit private college based in Dominica, but he and his students recently found themselves kicked off the island by a hurricane. As a result, he's teaching from a rented cruise ship elsewhere in the Caribbean.

I can't call or see him, Meisenberg says, but we can write to each other. And what follows is a long and candid email exchange.

"I can tell you about how modern races evolved," he tells me in his first message. It's clear from the outset that he believes he has an understanding of the subject not shared by mainstream scientists. The reason he is so happy to talk is because I'm giving him the opportunity to enlighten me. "One hang-up for academic definitions of race is that academics like precise definitions and precise boundaries between categories. They use only their left brain hemisphere. The right one is atrophied. For them, when categories into which they slice the world lack clear boundaries, they seem to assume that the categories are invalid."

On school performance in the United States, Meisenberg states, "Jews tend to do very well, Chinese and Japanese pretty well, and Blacks and Hispanics not so well. The differences are small, but the most parsimonious explanation is that much and perhaps most of this is caused by genes." There is no scientific evidence for this; it's just speculation. Nobody has ever found any genes linking ethnicity or race to school results. Like Henry Garrett half a century earlier, Meisenberg chooses to skip over the social, historical, and economic aspects of racial inequality. Rather, he believes that scientific evidence that doesn't yet exist will explain the gaps eventually. He takes it as given that the answer must be biological.

"Of course we can use molecular genetics to figure out, for example, in what way intelligence-related genetic variants vary among different racial groups," he suggests. "This will answer the question about race differences in intelligence once and for all. Good riddance [to] a stupid debate!"

He continues in this vein, with sensible statements punctuated by somewhat more bizarre ones. At one point he claims that "Europeans became brighter since antiquity, but then became stupider again since the 19th century." At another point he asks, "Are different races genetically predisposed to think about the world in slightly different ways?"

Eventually I ask what attracted Meisenberg—neither a geneticist nor a psychologist, but a biochemist working at a medical school—to this scientific niche. What got him so interested in race?

"I am not particularly obsessed with race," he replies. "But I got interested in the subject in the context of the question of why some countries are rich and others are poor." He notes that there is "a 50-fold difference in per capita GDP between the poorest countries in the world and the advanced Western countries" and he believes that learning ability, which he sees as being tightly linked to intelligence, is what makes the difference to a country's economic success. For him, this learning ability is programmed into a person's—or a population's—DNA. "In consequence, the question of whether there are genetic ability differences between people in different countries is perhaps the most fundamental question in development economics."

He tells me, for instance, what a shame it would be for what he calls "low-IQ countries" such as Pakistan to lose their brightest citizens if they emigrate to the West. "This cripples the poor countries and makes it impossible for them to catch up," he laments. If some nations don't have the cognitive ability to catch up to Europeans and East Asians, they will "get stuck somewhere on the lower rungs of the developmental path." At a stroke he condemns, without evidence, the whole world outside Europe and parts of Asia to genetically inferior status.

Meisenberg certainly seems bothered by the social implications of what he sees as immutable racial difference. He expresses a fear that the racial stock of smarter, wealthier nations is under threat and that this is a problem that urgently needs to be addressed, even if only by immigration control. "Populations that get too bright and too rich invariably slip into sub-replacement fertility and slowly breed themselves out of existence," he writes, "while those that are stuck at a lower economic and cognitive level also get stuck . . . with continuing high fertility of the less educated sections of the population." These could easily be the words of an early-twentieth-century eugenicist.

He, too, believes that the inferior might outbreed the superior.

| | | | | | | | |

The mystery for me as a journalist is not that scientific racists exist. There have always been those with prejudice of every persuasion in academia, and possibly there will always be. The bigger puzzle is how someone like Gerhard Meisenberg—a professor at a private university in the Caribbean—manages to keep the *Mankind Quarterly* afloat, and finds researchers to write the things that fill its pages. This requires networks, it requires coordination, and it requires funding. The fact is, across the world, old-style scientific race research of the kind his journal publishes is deliberately discouraged by science funding agencies and governments, not to mention deeply frowned upon within academia. It is as controversial as it was when the journal was founded. To do it independently, one needs resources.

Following the money trail is where the answers lie. Attached to one of his early emails, Meisenberg sends me a paper that attempts to describe how racial categories work. It was published in one of the 2017 editions of the *Mankind Quarterly*, authored by someone named John Fuerst, from an organization called the Ulster Institute for Social Research. I have heard of neither him nor it. When I do a check, the institute doesn't appear in the UK's list of officially recognized higher education bodies. It describes itself as "a think tank for the support of research on social issues and the publication of works by selected authors in this field."

It turns out that since 2015, the Ulster Institute for Social Research has been publishing the *Mankind Quarterly*, along with a handful of books on race. On the *Mankind Quarterly* website, its address is given as a postal box in London. Meisenberg himself sits on its advisory council. How the institute is funded he won't elaborate, revealing only that it operates on a shoestring budget. I find one link between the institute and an offshore company based in the Bahamas. "There certainly is no regular external funding from any outside source that I know of," he tells me. "I guess it's more a situation where someone may donate a larger sum, perhaps as part of a legacy. . . . That's how most of these small foundations work."

According to a report in the *Independent* newspaper in 1994, the institute received $50,000 the previous year from the Pioneer Fund. It was a period in which the fund, then based in Manhattan, New York, was particularly active. An investigation by the *Los Angeles Times* around the same time estimated that it was dispensing roughly a million dollars a year to academics, most of whom the newspaper claimed were looking for genetic differences between races. As well as giving grants to scientists, William Tucker has noted that between 1982 and 2000 almost $1.5 million was handed to lobbying groups favoring immigration reform in the United States.

If the Ulster Institute for Social Research really is operating on a shoestring today, part of the reason may be that the Pioneer Fund seems to have since declined to a standstill. "It is my strong sense that it is not nearly as influential as it once was, largely due to the deaths of the fund's key players," Tucker tells me. Staff at the Southern Poverty Law Center tell me that the Pioneer Fund has been fairly silent of late. As far as they can tell, in the last decade it has been gradually emptied of all its assets. In summer 2018 the Associated Press investigated the fund's tax records and found that it had disbursed nearly $7.8 million between 1998 and 2016.

Wherever their funds come from now, it is clear that there is a small cadre of researchers, some with very few academic credentials, who are still publishing and citing each other's research through organs such as the Ulster Institute for Social Research and the *Mankind Quarterly*. It is a closed, self-contained network. The same names crop up again and again. Richard Lynn, the assistant editor of the *Mankind Quarterly*, is also president of the Ulster Institute (one of its books is a tribute to him in his eightieth year). Lynn told the *Independent on Sunday* newspaper in 1990 that he had received grants from the Pioneer Fund. In 2001 he even published a history of the fund, titled *The Science of Human Diversity*. In his own published investigations, Robert Wald Sussman wrote that Lynn "does very little science and the 'science' he does is extremely poor."

Edward Dutton, of Oulu University in Finland, the author of at least two of the Ulster Institute's books, including one on racial difference in sporting ability, is also a regular contributor to the *Mankind Quarterly*. Another contributor, Tatu Vanhanen, a recently deceased Finnish political scientist, cowrote a book with Richard Lynn titled *IQ and the Wealth of Nations*, that appeared in 2002. Vanhanen was inspired to enter this area of research after reading up on evolutionary biology, which he interpreted as a way to explain interethnic conflict. He believed that political ideology could be used to serve people's genetic interests by keeping them loyal to their own ethnic group. In one high-profile magazine interview in 2004 Vanhanen claimed that the average IQ of Finns was 97, whereas in Africa it was between 60 and 70. The comment caused a national scandal because Vanhanen's son had just become prime minister of Finland.

Whatever the scientific merits of their work, the researchers in this group were and still are undeniably tight-knit. Most of them are largely unknown outside their circle but highly prolific within it. They have managed to build a thin veneer of scientific credibility that comes from getting published and cited, almost entirely by publishing and citing one another. And they keep finding new outlets for their work. The latest addition to this alliance is *Open Differential Psychology*, an open-access online journal that claims to have been set up in 2014 by a Danish research fellow at the Ulster Institute of Social Research called Emil Kirkegaard. It includes Gerhard Meisenberg and John Fuerst among its reviewers, and its published papers so far include a study of IQ in Sudan, and of crime among Dutch immigrant groups.

Even with the support of each other, those who write for the *Mankind Quarterly* rarely make much impact on science outside the shadowy recesses of the internet. But there have been a handful of higher-profile figures among them. Until his death in 2012, one was a Canadian psychologist, John Philippe Rushton, a former head of the Pioneer Fund and a professor at the University of Western Ontario. Rushton became notorious in academic circles for claiming that brain and genital size were inversely related, making black people better endowed but less intelligent than whites. Despite this, he was important enough to have his work read and reviewed by genuine scientists. One review in particular shone a spotlight on the kind of work that still routinely appears in the *Mankind Quarterly*. When Rushton's book *Race, Evolution, and Behavior* was published in 1994, the University of Washington psychologist David Barash was stirred to write, "Bad science and virulent racial prejudice drip like pus from nearly every page of this despicable book." Rushton seemed to have collected scraps of unreliable evidence in "the pious hope that by combining numerous little turds of variously tainted data, one can obtain a valuable result." The reality, Barash concluded, is that "the outcome is merely a larger than average pile of shit."

CHAPTER 5

# Race Realists

## *Making racism respectable again*

In 1985 Barry Mehler had a dream.

Now in his seventies and a history professor at Ferris State University in Michigan, where he studies genocide, in the mid-1980s Mehler had wandered into the murky territory of investigating the extreme-right-wing fringes of academia. His focus was on the founders of the *Mankind Quarterly* and Wickliffe Draper's notorious Pioneer Fund, both of which had been known to help keep fringe elements in science alive for decades. As Mehler worked, he found his waking life began to soak into his subconscious, coloring his sleep. In his dream—in truth, more of a nightmare—his son, then around two years old, was trapped in a runaway car hurtling down a hill towards oblivion.

"The traffic is going in both directions, and I am in the middle of the road desperately waving my hands trying to stop the flow of traffic in order to save the life of my son," he recalls. "It's a dream. It's a metaphor for how I felt."

For the previous few years, prompted by historical research he had already carried out into early-twentieth-century American eugenicists and their links to Nazi Germany, Mehler had begun looking into what happened to these same scientists and others with similar worldviews once World War II was over. Many people assumed that the eugenicists had all but disappeared with the Nazi regime, and that race science was pretty much finished at the same time. What Mehler learned instead was that the prejudice that had existed before the war—the fear of some kind of threat to the "white race"—was still alive in a few small intellectual circles.

"I was really focused on the ideological continuity between the old and the new, and the fact that these ideologies were malicious and dangerous," he explains. What worried him most of all as he did his investigations was that these people seemed to now be stepping outside their limited cabal, to penetrate not just mainstream academia but also politics. Their target was nothing short of the highest echelons of the United States government.

One of the key figures in the network to have survived from the old days was Roger Pearson, a founder of the *Mankind Quarterly* alongside Reginald Ruggles Gates. Pearson's career trajectory was very different from that of Gates. During World War II he had been an officer in the British Indian Army. In the 1950s he moved into working as the managing director of a group of tea gardens in what was then known as East Pakistan, now Bangladesh. And it was around then that he began publishing newsletters, printed in India, in which he explored issues to do with race, science, and immigration. Very quickly, says Mehler, Pearson connected with like-minded thinkers all over the world. "He really was beginning to organize, institutionally organize, the remnants of the prewar academic scholars who were doing work on eugenics and race. The war had disrupted all of their careers, and after the war they were trying to reestablish themselves. Establishing these institutional networks was essential for their rehabilitation."

Pearson's newsletters and the *Mankind Quarterly* relied on being able to reach out to marginal figures from all over the world, people whose views were generally unacceptable in the societies in which they lived. And of course this task had to be done before the benefit of the internet and social media, before it became easy for like-minded people with extreme views to find each other. "You have these people who seemingly come out of nowhere. It was just so amazing to me that they would be so well networked," says Mehler.

One of Pearson's publications was *Northlander*, which described itself as a monthly review of "pan-Nordic affairs," by which was meant, euphemistically, white northern Europeans more broadly. Its very first edition, in 1958, complained about the illegitimate children born as a result of "Negro" troops stationed in Germany after the war, and also about immigrants arriving from the West Indies into Britain. "Britain resounds to the sound and sight of primitive peoples and of jungle rhythms," Pearson warned. "Why cannot we see the rot that is taking place in Britain herself?" On the following page he printed a tribute to Charles Darwin. He had made it his goal to awaken people to what he saw as the existential threat of immigration and racial intermixing, referring to antiracists as "cosmopolitans."

Within a couple of decades, Pearson ended up in Washington, DC, setting up publications there, too, including the *Journal of Indo-European Studies* in 1973 and the *Journal of Social, Political, and Economic Studies* in 1975. "And that's what piqued my curiosity," Mehler continues. "Really looking at people who are racist at a time when liberalism was the predominant ideology." In April 1982 a letter even arrived for Pearson from the White House, bearing the signature of President Ronald Reagan, in which Reagan praised Pearson for promoting scholars who supported "a free enterprise economy, a firm and consistent foreign policy and a strong national defense." Somehow, he and those in his circle had managed to gain access to the very peak of the US government. An investigation published by the *Independent on Sunday* in 1990 confirmed that Pearson received several grants from the Pioneer Fund around the same time.

Just as Mehler was carrying out his research, a soft-spoken civil servant in Washington, DC, named Keith Hurt happened to be investigating the very same people in his spare time. When Mehler's and Hurt's paths crossed, the men began combining their research. Hurt was then working for the Congressional Research Service, a branch of the Library of Congress that provides policy analysis to members of the House and Senate, which meant that he was keen to keep his identity private. Today, he tells me, he is free to talk on the record. "I think I started out sort of naïvely," he admits. "I ran across some things that were disturbing, that I didn't expect. I didn't really understand that there were these structures and networks and associations of people that were attempting to keep alive a body of ideas that I had associated with at the very least the pre–civil rights movement in this country, and going back to the eugenics movement early in the last century. These ideas were still being developed and promulgated and promoted in discreet ways."

Perhaps paradoxically, given the fierce nationalism of the figures involved, this also appeared to be a global network, spanning the United States and Europe and also stretching to India and China. "If you looked at the old *Mankind Quarterly*, it was a truly international journal, with contributors and editors from all over the world," says Hurt. To this day, the *Mankind Quarterly* runs articles from many writers outside Europe, and its advisory board includes members in Russia, Japan, Saudi Arabia, and Egypt.

"It was important to put together how the networks worked, where the funding came from, what the publications were, what the connections were—and what the connection is with right-wing political organizations," adds Mehler. What he and Hurt were uncovering astounded them both. "There was a network, an international network of these people who were not particularly well respected or regarded or even known outside of the network, but they had their own journals, their own publishing houses. They could review and comment upon each other's work," explains Mehler. "So it was almost like discovering this whole little world inside academia. And it was a rather nefarious world, of people whose origins went back to the Second World War." What shocked them above all was the sheer professionalism of the operation, the slick ability of people with some of the most extreme views imaginable to connect to each other and communicate their views across thousands of miles. They were keeping scientific racism alive.

Mehler, who is Jewish, found it particularly disturbing. "I have a lot of relatives who survived the Holocaust," he tells me. "When they flip the light switch and the light goes on, for them it's like 'oh wow!' They are prepared for the world to collapse. They are prepared for things to cease to be normal very quickly because that was their experience." I can hear the fear in his voice, an anxiety that political stability in even the strongest democracies rests ultimately on a precipice. "I saw anti-Semitism. I was really alienated in American society. I was a person that felt that racism and anti-Semitism were predominant, and that the United States could easily become vicious, racist, and go back to its racist history when push came to shove, if people were threatened enough." The past, he reminds me, is always capable of repeating itself.

It was around this time, as he and Hurt uncovered the network, that Mehler had his dream. "I felt like I was desperately trying to prevent this from happening again. . . . I thought that we were headed for more genocide," he says. The parallels between this far-right network of pseudo-scientists and intellectuals and the rapid, devastating way in which eugenics research had been translated in Nazi Germany loomed large in Mehler's mind, terrifying him with the possibility that the brutal atrocities of the past could happen once more, that the ideological heart behind them was still beating.

| | | | | | | | |

Despite the urgency that both Barry Mehler and Keith Hurt felt, their investigations never made it into any high-profile publications. They appeared instead in a few small Jewish and left-wing newsletters, often with Hurt's name omitted, or an alias used, to protect his job at the Congressional Research Service. The lack of public interest reflected how many people assumed they no longer had anything to fear. Neo-Nazi political parties and white supremacists were thought to exist only on the irrelevant margins of real life.

"Race is such a difficult issue for Americans," explains Hurt. "People want to be optimistic. People want to believe that [racists] exist only in a sort of lunatic fringe, which is safely cabined off from the rest of society and that they have no consequences or implications for the future. That was the case then." The world was thought to be moving in a liberal, more inclusive direction. Racists were thugs and skinheads, not men in power, not academics, not covert networks.

Then, in May 1988, Mehler and Hurt published an article in *The Nation* that finally confirmed that there might be reason to worry after all. It linked a professor of educational psychology at the University of Northern Iowa, Ralph Scott to both the Pioneer Fund and the government. Scott had reportedly used his Pioneer grants under a pseudonym between 1976 and 1977 to organize a national antibusing campaign. According to Mehler, some of Scott's Pioneer Fund money had also been used to sponsor a study in Mississippi looking at the physical and psychological traits of "American Anglo-Saxon children."

What gave the story national significance was that, in 1985, the Reagan administration had appointed Scott the chair of the Iowa Advisory Commission on Civil Rights, a body with the express purpose of enforcing antidiscrimination legislation across the state. Just a few years earlier, Scott had brought, and later dropped, a lawsuit against three black civil rights activists who had described him as a racist. It was clear that as of 1985, Scott's views on race hadn't changed. Even after taking up his influential post, he continued to write pieces for the *Mankind Quarterly* and Pearson's *Journal of Social, Political, and Economic Studies*. Indeed his most recent article for the *Mankind Quarterly* was published in 2013. Scott, now an emeritus professor, refuses to give me any comments, or to confirm or deny Mehler's reports. But William Tucker has noted that almost every one of his papers was a variation on the same theme: that integrated schools were holding back white students, and not improving achievement among black students, for the simple reason that the two groups were somehow genetically different.

In short, here in 1985 was a man known to be actively involved in blocking policies aimed at achieving desegregation who had been made officially responsible for defending civil rights in his state. "It was obviously alarming," says Hurt. This also happened to come at a time when the Reagan administration was already facing criticism for drastically cutting the Civil Rights Commission's budget, making Scott's appointment look even more suspect. To outside observers it was hard to avoid speculating that Scott had taken up the position as chair of the commission to undermine it from within.

The month after Mehler and Hurt's article came out in *The Nation*, Ralph Scott resigned.

At the time, although the story made some corners of the national press, it wasn't major headline news. "It was largely dismissed, I would say, by people who in retrospect would probably admit were mistaken to write it off," says Hurt. Looking back on the case, in the context of today's politics, with the rise of far-right groups in Europe and the US, and nationalism more globally, he believes that what they uncovered should have served as a warning. Scott was just one individual, but he operated within a larger network of intellectuals opposed to desegregation, including Roger Pearson. "What surprised me was how quickly and efficiently these groups worked," adds Mehler. "You would think it would be fringe people, and that they would remain on the fringe, and they would have difficulty raising funds and making contacts. That wasn't true at all. What surprised me was how quickly Roger Pearson went from Calcutta, India, to Washington, DC, to Ronald Reagan."

Others within academia who have picked up the baton from Mehler and Hurt since then—including William Tucker at Rutgers University, who has done detailed investigations into the Pioneer Fund—have observed how well coordinated and resilient these networks have remained, even after key figures die. Tucker tells me that when he set out to research the Pioneer Fund and its wealthy founder, Wickliffe Draper, in the late 1990s, he "compiled a list of every academic or scientist I could think of who had been outspoken about racial differences and then searched the web or contacted their institution . . . to see if there was an archive of personal papers. In most cases there was. Then I traveled to each of these places, fully expecting that some trips would be a waste of time and research money, because there would not be any Pioneer connection." He was wrong. "In fact, I never struck out. Every one of these persons had been contacted and usually supported either by Pioneer or by Draper."

The Pioneer Fund may have since declined, but something important has happened to take its place. The global political landscape has changed, moving away from the center and making space once more for those at the extremes. The election of President Donald Trump, a Republican, has been joined by a rise in nationalist sentiment and far-right parties all over the world. For Hurt, the work he did three decades ago was prescient in relation to today's political climate, not because Roger Pearson or Ralph Scott were ever particularly important figures in US politics during Reagan's time, but because they managed to get close access to the government despite their views. Somehow, they both found a way to influence powerful people with their brand of intellectual racism.

"I've spent a lot of time in the last decade paying attention to the politics of immigration in this country, which are obviously related to all of that in intimate ways, and which dominate our politics in some ways today," Hurt tells me. He believes what happened in the past can happen again. "When Reagan came in, he didn't have established party networks of personnel that the establishment figures in the party had. So he cast a very wide net that included a very diverse range of people, including people like Ralph Scott. Scott wasn't, I think, representative of the sort of central policy thrust of the administration, but there was a lot of carelessness at the beginning of the Reagan administration that allowed these people to step into positions of greater or lesser significance . . . He was symptomatic of a broader problem of entryism to the Republican Party by people like this."

For those on the far right, it's a waiting game. As long as they can survive and maintain their networks, it's only a matter of time before politics swings around and provides an entry point once more. The public assumed that the extremes of scientific racism were dead, when in fact they were always active under the radar, says Hurt. "I think there was a whole sequence of events between the late 1980s and the present in which these ideas, which have become pretty well established in the mainstream of American political culture, were step by step progressing, reestablishing themselves, eroding the norms of the post–civil rights environment."

| | | | | | | | | |

What difference does it make to science that a publication such as the *Mankind Quarterly* exists today? In truth, barely any. Its work is so rarely read or cited by real scientists that its impact factor—the measure used to judge the influence of a journal—hovers between 0 and a little more than 1. By contrast, the impact factor of a highly respected journal such as *Nature* is more than 40. But then, of course, the *Mankind Quarterly* was never designed to be read by scientists or shape the future of research. It was always a platform for those looking for intellectual ballast for their political views. What is of concern, then, is what the journal represents. As a publication, it's a barometer for intellectual racism. Should it or its contributors become popular, then we know that something is wrong. And in the last ten years its impact factor has been on average higher than it was in the preceding decade.

At the same time its editors have built a presence in other more credible scientific journals. Assistant Editor Richard Lynn, for example, today sits on the editorial advisory board of *Personality and Individual Differences*, published by Elsevier, one of the world's largest scientific publishers, which counts the highly respected journals *The Lancet* and *Cell* among its titles. Among Lynn's papers was one in 2004, "The Intelligence of American Jews," in which he argued that "Jews have a higher average level of verbal intelligence than non-Jewish whites." Gerhard Meisenberg's work, which looks at the links between intelligence, genetics, and geography, has appeared in *Intelligence*, a psychology journal also published by Elsevier. He has published at least eight articles in recent years, including one in 2010 on the average IQ of sub-Saharan Africans, and another in 2013 on the relationship between "national intelligence" and economic success. When I check the editorial board members of *Intelligence* in 2017, I find both Meisenberg and Lynn listed.

Journals are free to publish whatever they think is worthy, subject to peer review. That said, the choice of whom to appoint to an editorial board is important because these members help to shape a journal's policy and scope. According to Elsevier's own online guidance, editors "should be appointed from key research institutes." Neither Lynn nor Meisenberg can claim that honor. A spokesperson for Elsevier, after repeated enquiries, tells me that editorial board members "are not involved in making decisions about which articles will be published. Their role is focused on reflecting the academic debate that takes place within the communities' domain that the journal serves." Yet Elsevier's own website states that editorial board members "review submitted manuscripts" and "attract new authors and submissions."

The other implication of their brief statement is that the work of Lynn and Meisenberg, studying population-level differences in intelligence—which some might reasonably equate with racial differences—must now be a part of mainstream academic debate.

In 2017 when I call the current editor-in-chief of *Intelligence*, Richard Haier, an emeritus professor at the University of California, Irvine, School of Medicine, to find out how he feels about having editors from the *Mankind Quarterly* on the editorial board of his own journal, he sounds nervous. "I struggled with this, frankly, when I became editor, and I consulted several people about this," he says. "I decided that it's better to deal with these things with sunlight and by inclusion." Throughout our conversation, he is uneasy, taking long pauses to choose his words. Keeping them inside the fold, he tells me finally, reflects his commitment to academic freedom.

Haier reassures me that he has never met Meisenberg or Lynn. But he tells me he did personally know and defend the late Arthur Jensen, a professor of educational psychology at the University of California, Berkeley. In 1969 Jensen mooted in the *Harvard Educational Review* that gaps in intelligence test results between black and white students might be because of genetics. It remains one of the most controversial psychology papers ever published. The *New York Times* reported in 1977 that the Pioneer Fund had been subsidizing Jensen's work. An investigation published by the *Los Angeles Times* almost two decades later, in 1994, confirmed that by then these grants to him must have totaled more than a million dollars.

Haier continues, "The area of the relationship between intelligence and group differences is probably the most incendiary area in the whole of psychology. And some of the people who work in that area have said incendiary things. . . . I have read some quotes, indirect quotes, that disturb me, but throwing people off an editorial board for expressing an opinion really kind of puts us in dicey area. I prefer to let the papers and the data speak for themselves." He adds, however, that he does believe there is something scientifically interesting about studying "group differences" in intelligence. "Scientific intelligence research has labored under this cloud for fifty years, and it is my stated goal as editor to help bring intelligence research back into the mainstream, where it used to be."

Even so, Haier seems to have been bothered by my questioning. When I check the Elsevier website around a year later, at the end of 2018, both Gerhard Meisenberg and Richard Lynn have been removed from the list of editorial board members of *Intelligence*.

If group-level or population-level differences in intelligence do need to come out from under the cloud of controversy, then what are the reasons that researchers might want to wade into this deeply divisive area of research? According to Richard Haier, one theme is shared by some of those who submit their work to *Intelligence*. "I can tell you, and I'm not revealing anything secret here as editor, we receive a number of papers that try to speak to the relationship between intelligence and economic development, and group differences in general." Sadly, it's not always of the highest quality, he admits. "When I read many of those papers, they are substandard, and they never even get to peer review. . . . We have had papers submitted that come up with some kind of result, and then the discussion section extrapolates from that result to immigration policy. Those papers never get as far as peer review in our journal because it's clear that there is an agenda." These were exactly the kind of extrapolations made by Gerhard Meisenberg in his emails to me.

Haier's comments betray just how much this field remains plagued by dark politics. I can't help but recall the nineteenth-century race scientists who jumped to biological explanations for the inequality they saw in the world, who thought other races had been doomed to failure by nature because their brains were too small or their temperaments too weak. Confronted by slavery and colonialism, they skimmed over history and culture, preferring instead to look to biology for justification for this kind of exploitation. When researchers like Meisenberg today link economic development to intelligence, they imply that the vast inequality between the world's richest and poorest countries is rooted not just in the imbalance of power or historical circumstance, but in the innate weaknesses of the populations themselves.

Racial injustice and inequality, in their minds, isn't injustice or inequality at all. It's there because the racial hierarchy is real.

| | | | | | | | | |

"I think what we're experiencing now is a much more threatening environment," says Keith Hurt. "We're in a much worse situation than we were a couple of decades ago." He believes that the kind of research once funded by the Pioneer Fund and still published by the *Mankind Quarterly* has now found fresh avenues of support. Scientific racism has come out of the shadows, at least partly because wider society has made room for it.

"Frankly, I think at this point the ideological stream that it was sustaining is now self-sustaining. There are other institutions, and a much, much broader culture that will sustain it."

This broader culture that Hurt describes, of people who rail against "political correctness" and call for a greater diversity of political opinion and freedom of speech in academia, has become stronger. In 2018 an investigation by the *London Student* newspaper revealed that the *Mankind Quarterly* editors Richard Lynn and Gerhard Meisenberg had been organizing and speaking at a series of small, invitation-only conferences held at University College London since 2014. The Associated Press reported in August 2018 that a University of Arizona psychology professor with an interest in human behavior and evolutionary psychology, and who had also been on the editorial advisory board of the *Mankind Quarterly*, had used a grant from the Pioneer Fund to attend one of the conferences. Somehow Lynn and Meisenberg had managed to secure a space within the university to discuss controversial issues around eugenics and intelligence, attracting one attendee, writer and commentator Toby Young, who was later appointed to head the UK government's Office for Students, the regulatory authority for the English higher education sector (he soon withdrew from the position).

Race scientists who had a platform for their views in the 1980s are building a stronger presence once more. In 1994, in *The Bell Curve*, one of the most notorious bestsellers of the twentieth century, political scientist Charles Murray and psychologist Richard Herrnstein suggested that black Americans were less intelligent than whites and Asians. A review at the time in the *New York Review of Books* observed that they cited five articles from the *Mankind Quarterly*, and no fewer than seventeen researchers who had contributed to that journal. Murray and Herrnstein went so far as to describe Richard Lynn as "a leading scholar of racial and ethnic differences." Although *The Bell Curve* was widely panned after it was published—an article in *American Behavioral Scientist* described it as "fascist ideology"—in 2017 *Scientific American* magazine noted that Charles Murray was enjoying "an unfortunate resurgence." Facing down protestors, he was being invited to give lectures on college campuses across the United States.

Another contributor to the *Mankind Quarterly* has become a key figure in the white supremacist movement. Yale-educated Jared Taylor, who belongs to a number of right-wing groups and think tanks, founded the magazine *American Renaissance* in 1990. William Tucker calls the magazine the true intellectual arm of the modern neo-Nazi movement. For example, Taylor uses a concept to defend racial segregation that he borrowed from the zoologist Raymond Hall, writing in the first ever issue of the *Mankind Quarterly*: "Two subspecies of the same species do not occur in the same geographic area." His brand of white supremacy draws from race science to lend itself the illusion of intellectual backbone. He is in some ways the Wickliffe Draper of the twenty-first century.

Like Draper, Taylor has sought to make racism respectable again. Robert Wald Sussman has described his American Renaissance Foundation conferences as "a gathering place for white supremacists, white nationalists, white separatists, neo-Nazis, Ku Klux Klan members, Holocaust deniers, and eugenicists." The *Journal of Blacks in Higher Education* has more succinctly dubbed them "a convocation of bigots." The conferences have featured other intellectuals who have written for or edited the *Mankind Quarterly*. A visitor at the 1994 meeting reported that people didn't "flinch from using terms such as 'nigger' and 'chink.'" Male attendees are expected to dress in smart business suits, to set themselves apart from the thuggish image most people associate with racists. They don't call themselves racists. They call themselves "race realists," a euphemism that reflects how they like to believe the scientific facts are on their side.

By the time of the 2016 US presidential election, Jared Taylor's place in American politics was firmer than ever. He even appeared in a television ad released by Hilary Clinton's campaign team to show the kind of support that her rival Donald Trump's anti-immigration policies had among white nationalists. Explaining the rising prominence of people such as Taylor during the campaign, one correspondent reflected in *Newsweek* magazine in 2017: "These men have degrees from some of the nation's top universities. . . . They are a well-read group who cloak their ideas about the intrinsic superiority of white men in selected passages of literature, history, philosophy and science."

One regular speaker at American Renaissance Foundation conferences who is also a *Mankind Quarterly* contributor is Michael Levin, a professor of philosophy at the City University of New York, which describes itself as a large, relatively affordable urban university that is committed to being accessible to underrepresented minorities. He is the author of *Why Race Matters*, a book that went out of print after its initial publication in 1997, but was reissued in 2016 with a new cover and a foreword by Jared Taylor. In 1986 Levin had written a letter to the *New York Times* arguing that it was legitimate for store owners to discriminate against all black people because they were more likely to be attacked by someone who was black. According to the

Southern Policy Law Center, Levin told the audience at a 1998 American Renaissance Foundation meeting, "The two principal race differences that I see are race differences in intelligence and in motivation. . . . It's no wonder there are very few black scientists. . . . You have to have an IQ of 130 to be a successful research scientist." According to an article in the *Journal of Blacks in Higher Education* in 1994, Levin had by then received more than $120,000 in grants from the Pioneer Fund.

"People like that are able to get funding," Keith Hurt tells me. "There will always be, unfortunately, men of wealth—and they are almost all *men* of wealth—who share these ideas and are willing to support them." But another reason that scientific racists have more influence now is that the internet and social media have given them simpler ways to access and grow their networks. "Behind every racist joke is a scientific fact," the alt-right blogger Milo Yiannopoulos told a *Bloomberg* reporter in 2016. Among the cabal of online race realists and their followers it's easy to spot an attitude of doggedness. They repeatedly insist that they are challenging the politically correct wider world by standing up for good science and that those who oppose them are irrational science deniers.

Gerhard Meisenberg, for example, writes to me in our correspondence that "some academics seem to believe that by simply claiming that race differences don't exist, we can prevent people from believing in them. Doesn't work like that. . . . For example, if we tell people that black children are as smart as white children and it isn't true, there will be teachers and others who know first-hand that it isn't true. Also, if we tell people that it's because of some flaws in the school system and the flaws are repaired and it doesn't help one bit, not only do people get frustrated about all the wasted effort, but they also start distrusting the 'scientists' who are telling them lies. Real people aren't postmodernists. They distinguish between truth and lies." Meisenberg's tactic is simple: he uses people's gut prejudices and casual stereotypes to undermine trust in mainstream science. If you feel it to be true, it must be.

This rhetoric around who has the genuine claim on the truth resonates today more than ever when the public on both sides of the political divide worry about fake news and media conspiracies. Yet it's a line that has been adopted by intellectual racists for decades. In 1998 an American corporate lawyer, Harry Weyher, president of the Pioneer Fund from 1958 until he died in 2002, was given space to write an eighteen-page editorial for the journal *Intelligence*. He used the opportunity to defend research supported by the fund. "This is critical research by world-class scholars. . . . Yet, if one were to believe some important segments of the media, this research was funded by an evil foundation and done by evil scientists, and is unfit for public dissemination."

Twenty years before Donald Trump was elected, Weyher, too, laid into what he saw as an egalitarian orthodoxy and political correctness, going so far as to accuse the print and broadcast media of "false reporting."

| | | | | | | | |

"Why do we still have race science given everything that happened in the twentieth century?" I'm asked by Jonathan Marks, an anthropologist who is the academic many turn to for clarity when it comes to racism in science.

His answer is unequivocal: "Because it is an important political issue. And there are powerful forces on the right that fund research into studying human differences with the goal of establishing those differences as a basis of inequalities." Ultimately, politics is always a feature of the science, just as it was in the very beginning. Once there was the backdrop of slavery and colonialism, then it was immigration and segregation, and now it is the right-wing agenda of this age. Nativism remains an issue, but there is also a backlash against greater efforts to promote racial equality in multicultural societies. And just like before, the message of those with racist intentions is tailored gently, carefully sculpted to appeal to populist fears while at the same time sounding logical and reasonable.

For example, when communicating with me, Gerhard Meisenberg uses the word "culture" alongside the word "race," as though they're fully interchangeable. He fully grasps that most people these days value and respect cultural boundaries, even if they don't recognize biological race—so he seeks to conflate race and culture.

"Without much selectivity in migration, all countries of the world become homogenized, not only in ability level but also in culture and everything else. Countries become more similar to each other," writes Meisenberg. How tragic it would be to have the whole world look exactly the same. On a logical level he fails to explain how, if races are fixed and immutable in the way he thinks they are, we could all end up the same just by migrating. But this isn't the point. On the surface, heard quickly, his concerns sound almost sensible, in the same way that eugenics sounded so logical and attractive to progressive liberals in the early twentieth century. Who back then could argue against the pursuit of a healthier, stronger population? Who today could argue against countries and groups maintaining their distinct cultures?

Race has always been an intrinsically political area of research, the idea itself born out of a certain world order. So it's small surprise that those doing this kind of research end up in the same place again and again, as they have for centuries. When they look for human variation, however objective they claim to be, they can't help but ask what the differences they think they see mean for society. This is why race science so often comes twinned with speculation on economic and social differences.

A common theme among today's race realists is their belief that, because racial differences exist, diversity and equal opportunity programs—designed to make society fairer—are doomed to fail. "As far as I understand, race-based policies of this kind were adopted by many American institutions since the 1960s or 70s, and were originally justified as something that is needed to make up for disadvantages that Blacks and some other minorities had suffered in the era before civil rights legislation," Meisenberg explains. "Today, 50 years later, that old reasoning is no longer credible." Rather than investing in these policies, he appears to be arguing that we should accept inequality as a biological fact. If an equal world isn't being forged fast enough, the race realists don't see it as a longer or tougher path than we imagined and so we should redouble our efforts. Instead, it is a permanent natural roadblock created by the fact that, deep down, we're not the same.

"We have two nested fallacies here," says Marks. The first is that the human species comes packaged up in a small number of discrete races, each with its own different traits. "Second is the idea that there are innate explanations for political and economic inequality. And basically what you're doing there is saying that inequality exists, but it doesn't represent historical injustice. What these guys are trying to do is manipulate science to construct imaginary boundaries to social progress."

For Marks, the prescription to cure this pathology is radical. Banning any kind of scientific research—if it can get funded and ethically approved—is a risk to academic freedom. He suggests instead that people who "cannot handle the results shouldn't be studying it. . . . We don't want racists working on human variation because that doesn't work. So it's not a question of 'should this be studied?'; it's a question of who should be studying it and how should they be credentialed, or how should they be vetted." His argument is that if this is a field that is always going to be affected by the politics of the scientists, then surely it makes sense to have people doing the research who aren't bent on division and destruction, whose aims don't lie at odds with what society as a whole has decided is morally acceptable.

To others, this sounds heavy handed. For example, William Tucker, the psychologist whose painstaking work helped expose the Pioneer Fund and the scientists it has backed, supports the freedom of anyone to do any research—even the kind he abhors. "People enjoy the right to take Pioneer's money; I would not like to see them deprived of that right. At the same time, I think that their decision to do so is awful and will do whatever is in my power to persuade them of the folly of such a course," he tells me.

Fundamentally, though, the problem is not the science itself. If it were limited to academia, the kind of material that's published in the *Mankind Quarterly* and other like-minded publications would have next to no impact at all because mainstream scientists almost entirely ignore it. It just doesn't have enough scientific value. As for the ideology, most researchers today accept that what little we know about human variation can't be used to dictate how we treat people in the real world. It certainly can't be used to set policy. The problem is in how these ideas are used and abused in the wider society, how much traction they can get with the public and those in power. Nazi scientists carrying out their regime's program of "racial hygiene" had only a rickety scientific scaffolding upon which they wreaked enormous destruction on millions of people. The same was true of those who defended slavery, colonialism, and segregation. And it applies today, too, to those on the political extremes. Nothing is more seductive that a nice string of data, a single bell curve, or a seemingly peer-reviewed scientific study. After all, it can't be racist if it is a "fact."

For those with a political ideology to sell, the science (such as it is) becomes a prop. The data itself doesn't matter so much as how it can be spun. Marks warns me that those to really watch out for are the ones who claim to be uniquely free of bias, who tell you they have a special, impartial claim on the truth. "Whenever anybody tells you 'I am objective, I am apolitical,' that is the time to watch your wallet, because you're about to have your pocket picked."

And he should know, because he almost had his pocket picked.

CHAPTER 6

# Human Biodiversity

## *How race was rebranded for the twenty-first century*

It was 1998 or thereabouts that an invitation arrived in Jonathan Marks's email inbox. He can't recall the precise date because until I asked him, he hadn't thought about it for years. What he does remember is that the sender was a little-known science journalist and former writer for the conservative *National Review*, Steve Sailer. The invitation was for Marks to join a mailing list of people interested in the subject of human variation.

"I knew absolutely nothing about him," Marks recalls. "He just seemed to be someone who was organizing something."

At the time, Marks was teaching at the University of California, Berkeley, a few years on from having written a popular textbook on race, genes, and culture, titled *Human Biodiversity*. Wedging these two words together, he had neatly coined a phrase to describe biological and social variation across the human race. Part of the reason he chose it, he says, is that "biodiversity" had become something of a buzzword. He never guessed it would cause any problems. And why would it? Diversity was being celebrated; both the glorious biodiversity of the natural world and the rainbow of cultural and physical diversity in human societies. It was the proud label of liberal antiracists, of the good guys like himself.

"America's answer to the intolerant man is diversity," Robert F. Kennedy affirmed in 1964 at the dedication of an interfaith chapel in Georgia. "United in diversity" was the motto of the European Union. It goes without saying that although different cultures have different qualities to offer, the breadth of human variation does nothing to undermine the general consensus that we're biologically pretty much all the same beneath the skin. This is a truth that's been universally acknowledged since the end of World War II.

At least that's what Marks thought.

The invitation itself appeared perfectly innocent. The 1990s marked the early years of electronic mailing lists, and Steve Sailer apparently wanted to use one as a way of pulling together scientists, intellectuals, and fellow journalists to start a private conversation about human difference. "He said, 'Hey, I'm interested in human variation, and I like your work. Let's get an email list together,'" says Marks. "It seemed pretty straightforward and harmless." So he signed up. What intrigued him especially was that Sailer happened to be brandishing Marks's own neologism, calling his list the "Human Biodiversity Discussion Group."

Others joined in their dozens. By the summer of 1999, Sailer's roster of members was astounding. Along with prominent anthropologists such as Marks, there was psychologist Steven Pinker, political scientist Francis Fukuyama, and economist Paul Krugman. In hindsight, the large number of economists in the group might have been a warning. There in the mix, too, was the controversial author of *The Bell Curve*, political scientist Charles Murray. That should have been another red flag.

When Marks had talked about human biodiversity, he meant the superficial variations we see among individuals, right across the species—not variation among human groups. "There isn't really room for three, or four, or five biologically distinct kinds of people," he tells me. He certainly didn't expect to see people on the list reinforcing old-fashioned racial stereotypes of the kind that had long been debunked. That school of racism was long dead, he assumed. Yet here on this email list, something strange was happening. Observing the conversations that Sailer steered through the group, Marks noticed the term "human biodiversity" being used differently from the way he had originally intended. Members were using it to refer to deep differences between human population groups.

Among the people added to the group that summer was Ron Unz, a Harvard graduate and founder of a financial services software company who had recently run as the Republican candidate for the governor of California. His introduction to other members was pasted alongside a 1994 article Unz had written in the *Wall Street Journal*, "How To Grab the Immigration Issue," on the changing demographics in California's politics. "Conservatives and Anglos have become enormously angry and frustrated over the growth of crime, welfare, affirmative action and the general decay of their society," he stated. Another addition to the group was Deepak Lal, a professor of international development at the University of California, Los Angeles, whose work explored the reasons the West was economically more successful than the rest of the world.

It dawned on Marks that Sailer's seemingly innocent email list was not so much a way to discuss science in an objective way but more about tying together fresh science and economics with existing racial stereotypes. One debate that sticks in Marks's mind today, for instance, took place between him and a journalist who claimed that black people were genetically endowed to be better at sports. Marks insisted that this was a scientifically shaky argument, not to mention one with dangerous political implications. The two experts clearly disagreed. But rather than help reach a consensus, "Steve Sailer clearly took his side," he tells me.

"At which point I realized, 'Ah! This isn't an impartial scholarly discussion.'"

Another time, Sailer defended a writer on the list who suggested that different ethnic groups had their own particular strengths—for example, saying that Turks were born physically more powerful than other groups. This was, Sailer suggested, one reason why affirmative action policies to hasten racial equality weren't a good idea. Different ethnicities should instead be encouraged to do what they do best. When Sailer talked about human biodiversity, he didn't appear to be using the phrase in a politically neutral way, but as a euphemism. He had spun the language used by liberal antiracists to celebrate human cultural diversity to build a new and ostensibly more acceptable language around racism.

In email correspondence with me, Sailer denies duping anyone, although he does admit that the group leaned toward the heretical.

Whatever his intentions, this wouldn't have been the first time someone had tried to subvert the idea of diversity. In their conclusion to *The Bell Curve*—a book which just a few years earlier had notoriously claimed that black Americans were innately less intelligent than whites—Charles Murray and Richard Herrnstein similarly undermined the political push towards racial equality by arguing that biological differences between groups made racial equality practically impossible. "We are enthusiastic about diversity—the rich, unending diversity that free human beings generate as a matter of course, not the imposed diversity of group quotas," they wrote. Every person in a diverse society had a valued place, they implied, just not the same place.

Realizing his error, Marks left the group. "If somebody said the same to me today I would probably be a little bit more suspicious," Marks admits. "I would look at who else is on the invite list. But even that can be misleading because he may be inviting people who are just as confused about this as I am." At the time it seemed harmless. Just a bunch of people with some marginal political ideas trying to convince others of the same.

Instead, Marks was about to find out just how prophetic the existence of the list was. "That was my introduction to what became the alt-right."

| | | | | | | | | |

For those sucked into Sailer's electronic arena for the intellectual discussion of race, his email list was just a taste of the virulent racism that would later be seen far more often in the shadowy areas of the internet, then more openly on social media and right-wing websites, and finally in mainstream political discourse. Many more soon took hold of the phrase "human biodiversity," giving it a life of its own online. Today it's nothing short of a mantra among self-styled race realists.

"Blogs have made the dissemination of wacko ideas much more efficient," says Marks. "Actually, in academia, we don't really know what to do with blogs because they have a really short shelf life. You forget about them a day after you read them; it's hard to cite them. But today, they are out there. They're these markers of people with wacko ideas."

To be fair, few could have guessed that the email list was a precursor to something bigger. But as the group slowly went defunct, Steve Sailer's political convictions became increasingly obvious. He and other members of the list went on to become prominent conservative bloggers, writing frequently on race, genetics, and intelligence. As a columnist for American website VDARE.com, which describes itself as a news outlet for patriotic immigration reform (and, like the *Mankind Quarterly*, is automatically blocked by my internet service provider), Sailer once argued passionately for the biological reality of race, stating that it was a fundamental aspect of the human condition. In 2009 the same right-wing website published Sailer's first book, about Barack Obama, titled *America's Half-Blood Prince*. In 2013 the software mogul Ron Unz founded his own blogging platform, the *Unz Review*, as an alternative to the mainstream media; he recruited Sailer as one of his most prolific columnists.

Sailer rose to true prominence, though, in the US presidential election of 2016. Six years earlier he had proposed focusing heavily on immigration to draw in white working-class voters as a single bloc. Having subverted the language of diversity through his email list, he did the same with identity politics. If ethnic minorities, such as black and Hispanic Americans, could assert their rights and defend their interests, he reasoned, then why not white voters who felt they were losing out to cheap immigrant labor and globalization? It turned out to be an unexpectedly successful campaign strategy for Donald Trump. Dog-whistle politics was reframed as a pushback to liberal elitism.

"I think the big problem this country has is being politically correct," Trump said in the first Republican Party presidential debate, in Ohio in 2015. A similar approach was adopted by some campaigners in the UK in favor of leaving Europe during the Brexit debate around the same time.

An article in *New York* magazine in 2017 described Sailer's string of prophetic political insights as a new wave of populist thinking on the right, dubbing it "Sailerism." In the United States at least, he was credited with inventing a form of identity politics for disgruntled poorer whites, a group that had been neglected by politicians. But at the same time, political observers couldn't help but notice that his ideology often looked suspiciously like white nationalism.

For Keith Hurt, in Washington, DC, who investigated right-wing intellectual racism in the 1980s and has kept a close eye on the immigration debate ever since, none of what happened during that time should have come as a surprise. "The election of Trump made it impossible for many people to any longer overlook this stuff," he tells me.

Hurt explains that the racist ideologies that existed at the start of the twentieth century, which manifested themselves in the eugenics movement and then German nationalism, survived to the end of the century. The only difference was that those who held these views were later forced to keep them private. "The post–World War II ideological consensus sort of rested on an implicit social contract that said, on the one hand, overt racist language will not be deployed in the public square in the way it was before. On the other hand, if this kind of talk is pushed out of the public square, society will refrain from making accusations of racism against all but the most extreme fringe," he says. And so the intellectual racists, the ideologues, communicated with each other through their own tight networks and disseminated their ideas through their own publications, some of which were so marginal and private that they were almost invisible to the outside world. It was easy for the public to assume that they didn't exist, that the only genuine racists left were the ones they could see and hear, the skinheads and thugs.

When the time was right, however, political and intellectual racism slowly resurfaced in the mainstream. Sensing the rising tide, people like Steve Sailer waded in. "They couldn't exist as they do now if there hadn't been the intellectual ideological continuity," adds Hurt. They weren't starting from scratch; they were just repackaging the old ideas, the existing racial assumptions that had been around for many decades.

But it all came as more of a surprise to academics like Jonathan Marks. "I was working on the assumption that these guys were a lunatic fringe. If you had told me twenty years later that they would be part of a political mainstream wave, I would have said you are absolutely crazy. These guys are antiscience. These guys are positioning themselves against the empirical study of human variation and they are clearly ideologues for whom empirical evidence isn't important," he says with a laugh. "But I think they were a lot cleverer than us professors."

| | | | | | | | |

Offline, in mainstream academia and among respected scientists, a different debate was taking place—one that would, similarly, come to reshape the way people thought about race.

It was 1991, shortly after the launch of the multi-billion-dollar Human Genome Project, which was steadily working to build a map of the genetic data shared by our species for the first time by sequencing the entire length of human DNA. At the Stanford University School of Medicine, an influential geneticist by the name of Luigi Luca Cavalli-Sforza, then around seventy years old, spotted an opportunity. His own long and illustrious career had been dedicated to studying human variation across the world, work that had taken him from Italy to the United States and brought him to the forefront of a field known as population genetics. Cavalli-Sforza and some of his colleagues—a small team of anthropologists and geneticists, based mostly in the United States—wondered whether the same kind of data being collected by the Human Genome Project could also be used to pick through the fundamental differences between human groups, the genetic variations that make us who we are. Little did he guess that it would turn out to be one of the most controversial scientific initiatives of its time, leading to decades of debate about the true biological reality of race.

Unlike Steve Sailer, Cavalli-Sforza and his team didn't come to their new project with race in mind. Indeed, quite the opposite. They were avowed antiracists, fully signed up to the UNESCO statements on race in the 1950s, and firm believers that science would do nothing but prove racial stereotypes incorrect. They thought that their work might even help free the world from the scourge of racism. "This was, at least on the American side, in general political terms, a quite left-wing group of people," recalls Henry Greely, a professor at Stanford Law School who became involved in the initiative later on.

Cavalli-Sforza himself was used to receiving hate mail from people who didn't agree with his outspoken belief that genetics didn't support old-fashioned notions of race. In 1973 he had publicly debated William Shockley, a Stanford University physicist and joint Nobel Prize winner, who in later life became a notorious racist. Shockley believed that black Americans had intellectual shortcomings that were hereditary, and that black women should therefore be voluntarily sterilized. He was among the most prominent race theorists to receive support from the Pioneer Fund. When they met for a debate at Stanford, Cavalli-Sforza coolly demolished Shockley's claims, fact by fact.

The field of population genetics itself was also born of post–World War II efforts to move away from traditional race science and eugenics. In the 1950s and 60s, when geneticists stopped talking about race, they turned their attention instead to "populations," the "human variation" that existed between these populations, and the "frequencies" of alleles—different versions of the same gene—that existed within these groups. It was a more rigorous, mathematical, molecular approach to studying human difference. In the course of this study, population geneticists such as Cavalli-Sforza quickly noticed that there were no hard genetic boundaries around human groups, but rather continuous statistical variation, with a good deal of overlap. What differences there are exist along gradients, not borders.

That said, variation isn't random either. Depending on how you look at it, it can fall into clusters in which certain genes are statistically more common in some groups than in others. Population geneticists became fascinated by these clusters. They assumed that they would be more obvious in places where people had been geographically isolated for hundreds of years, living on islands or at the tops of mountains, mating almost exclusively within their own group through many generations. "Primitive groups" were believed to be particularly distinct genetically because of the length of time they had spent away from others, shielding their genomes from the effects of intermixing. Being so remote, their lineage made conspicuous thanks to their long isolation, maybe their genes could offer special insights into how humans evolved and adapted.

Cavalli-Sforza was among those who believed that by studying the genomes of these primitive groups it might be possible to track historic patterns of migration. If they could see how gene frequencies varied in them and their closest neighbors, they could perhaps track where their ancestors once lived in the distant past. In 1961 he became one of the first scientists to apply modern statistical methods to see how frequencies of major human blood types varied among large human groups; he created a "family tree" of blood types to show how these groups were related.

The search for what they thought would be genuine difference at the far reaches of the planet took researchers to the most remote places. In 1964 the World Health Organization picked out Eskimos in the Arctic, Guayaki hunter-gatherers in the forests of eastern Paraguay, pygmies in the Central African Republic, Aboriginal Australians, and the tribes of the Andaman Islands in India as potentially interesting "primitive groups." Scientists who couldn't make the trips to see them would settle on studying immigrant groups in their own backyards, particularly Jewish and Roma communities, also known to be ancient and tight-knit. There was no point in studying very large human populations—say, just Africans or Europeans—because there was too much variation within them as a result of migration and intermixing. Small, old indigenous communities were thought to be special, and more distinct.

On the back of all this research, Cavalli-Sforza formed a new plan. He mooted the idea of using the same revolutionary gene sequencing technology as the Human Genome Project to map not just one genome, but to travel the world and draw genetic data from lots of individuals of different ethnicities. This Human Genome Diversity Project, he and his collaborators announced in the journal *Genomics* in 1991, would "supplement and strengthen findings from archaeology, linguistics, and history." Through their work, genetics would take its place alongside the study of culture, language, and history to help paint the whole human story.

The Human Genome Diversity Project wanted initially to zoom in on four hundred to five hundred different small populations, especially geographically remote and apparently dwindling indigenous ones—including the Basques in Europe, the Kurds of eastern Turkey, and Native Americans—which they referred to as "isolates." They expected the groups' genetic data to provide clear patterns that might shed light on prehistoric migration and social structure. Calling for urgent funding from the world's governments, their main argument was that this was a race against time to document the breadth of human variety before we all entered the great big melting pot, when migration and assimilation would leave us all so genetically similar that such an effort might not be quite so worthwhile.

But what they were proposing didn't sit easily with everyone. Some outside observers couldn't help but be a little uncomfortable. After all, it was hard not to wonder whether in the nineteenth century, this might have just been called race science. The Human Genome Diversity Project was technically more precise, of course, more scientific. It wasn't sampling skin and hair color, or slotting people into racial hierarchies. It was using genetics. But in some ways it was hardly distinguishable from the study of human difference a hundred years earlier. The word "race" had been prudently replaced by "population," and "racial difference" by "human variation," but didn't it look suspiciously like the same old creature?

Then again, could it be called race science if the scientists involved were obviously antiracists? As Henry Greely told me, they were left-wing liberals committed to stopping racism, whose public lives had been dedicated to fighting scientific racists and eugenicists. How could there be anything to fear?

| | | | | | | | | |

"They won't use the term 'race,'" says the Yale University historian Joanna Radin, who has studied the progress of the Human Genome Diversity Project from 1991 to the present day. She notes that there was always a deliberate effort to avoid the word "race," mostly because of its obvious political baggage but also because the scientists didn't see genetic "isolates" as being "races" in the traditional sense. They were different enough to merit study, but not in the way racial difference had been described in the past. They were small groups, not large continental-scale ones. "They don't necessarily map onto existing racial hierarchies."

Those behind the project insisted that their research was countering racial myths. Their stated intention was to replace ignorance and prejudice with hard scientific facts, and make it clear that we are one single human species, united in our common origins. Their plan was not to look for difference as racists had in the nineteenth century, to prove inferiority or superiority, but to use the tiny difference there deep in our genomes to help build a picture of human migration. They would be like archaeologists, digging through our genes, looking for clues about our history. Their aim was simply to understand our past.

Given the unimpeachable political credentials of scientists such as Luigi Luca Cavalli-Sforza, the scientists behind the Human Genome Diversity Project might have expected it to go forward without a hitch. All they needed was funding, and permission from their "isolates," the indigenous communities whose blood and DNA they wanted to sample.

But things had become more complicated. This wasn't the sixties anymore, when foreign researchers could pick the communities they wanted to study and just expect them to comply. They had become more wary. "Luca Cavalli-Sforza was an old school anthropologist," geneticist Mark Jobling at the University of Leicester explains. "I mean, I went to talks by him in the nineties where he would show old slides of him collecting DNA, blood samples in Africa from Pygmy groups, and offering glass beads and cigarettes in return, things like that." This wasn't the way things were done now. This was the 1990s, the dawn of the internet age, identity politics, and the fight for indigenous rights.

Scientists who took an interest in indigenous communities found that now, the same communities were taking an interest right back. They weren't as trusting as they had been in the past. And they were organized. And they had good reason to be organized. Remote tribes and ethnic groups had been exploited throughout history, their land and cultural artifacts pillaged by Western colonizers, their bodies targets for unethical experimentation. Between 1946 and 1948, for example, the United States government ran secret experiments on thousands of people in Guatemala in which they deliberately exposed them to sexually transmitted diseases. Before and during World War II, British scientists deliberately sent Indian soldiers fighting for Britain into gas chambers to study the soldiers' response to exposure to mustard gas. There was a long and bloody tradition of scientists abusing other populations for their own ends, particularly those populations deemed at the time to be racially inferior. But people were now prepared. Rights activists, alert to the risk of exploitation, were ready to defend the communities targeted by the Human Genome Diversity Project.

These activists warned of the possibility that DNA analysis might damage how these communities chose to understand their past, might reveal something that could be used commercially to extract profit, or even be used as a weapon by racists. They weren't prepared to hand over their biological data, their blood and tissue, knowing that there was a possibility that it could end up misused.

Even so, their resistance to join in the project baffled some of the scientists involved. As Yale's Joanna Radin explains, "What had changed—and this is what caught the scientists by surprise—was that the indigenous groups they had imagined disappearing, that they didn't have to reckon with once they left, these purported isolates, were organizing in indigenous movements. They had access to the web; they were in touch with activists."

As the future of the project came increasingly under threat, Cavalli-Sforza and his colleagues called in Henry Greely from Stanford Law School to help navigate the ethical dilemmas and deal with critics. Greely, fascinated by the project and taking a personal liking to the participants, agreed. At the beginning, he knew very little about the science itself. "I knew how to spell DNA but that was about it." But from the outset, he could see that there were likely to be problems. The scientists were of course aware that science did not have a great historical record when it came to race. "They knew this was an issue," he says. "They knew that it had a bad past." But they didn't see themselves as part of that past. As far as they saw it, "They were the *break* from the past. They were the good guys. They were the ones who understood the rights-enhancing, equality-enhancing potential of genomics, and they were going to bring it to the world." Greely adds that this was the first place he had heard "that all humanity is more similar to each other than a band of chimpanzees that lives in a particular region of Africa. They expected to see that people were quite similar."

But at the same time, they were oddly naïve about how the project might be seen from the outside. One scientist "talked about the need to sample 'isolates of historical interest,' a term that indigenous populations did not care for," Greely admits. "It struck me that that was not likely to be well received because it's a very clinical, bloodless way of referring to people who are alive, and cultures that are living now. Historical interest is sort of something you find in a museum. It was tone-deaf. Naïveté is always easily diagnosed through the retrospectoscope."

Greely's job, to navigate the ethical quandaries posed by the Human Genome Diversity Project, turned out to be a poisoned chalice. Partly because of the political controversy surrounding the project, it didn't attract the funding they wanted. Yet, throughout, the scientists struggled to understand why. "Some were so comfortable with their own knowledge of their own moral bona fides that it was hard for them to imagine being attacked from the left," says Greely. "They would have imagined that any remaining *racists* would be attacking them, not that *they* would be attacked as being racists." If anything, the scientists were on the same side as those attacking them. "I was the most conservative person on the North American committee, and I'm a Carter-Clinton-Obama Democrat!"

The activists representing indigenous groups turned up the dial on their protests. At a meeting of the World Council of Indigenous Peoples in Guatemala in 1993, Greely even found himself facing down the untrue accusation that he was a CIA agent, intent on committing genocide. "We took some lumps. Some of them were deserved and some of them weren't."

In an address at a special meeting of UNESCO in 1994, Cavalli-Sforza turned his focus to the charge of racism, insisting that his project would help combat prejudice, not perpetuate it. But in 1995 another political storm was created when scientists funded by the United States National Institutes of Health tried to patent a virus-infected cell line from people belonging to the Hagahai tribe in the highlands of Papua New Guinea for the purpose of developing a new treatment for leukemia. Activists accused them of stealing people's biological samples and attempting to profit from them.

It was in this charged political environment that Greely wisely drew up a model ethical protocol, setting in stone that samples wouldn't be taken unless entire groups, not just individuals, had given their consent. For scientists, it seemed obvious that they were not ill intentioned.

But if people were slow to see their good intentions, there was another, deeper reason. At the same time that the project claimed to be antiracist, it was hard to escape the paradox that this was also all about finding out how people differed. If the genetic variation between us was already known to be trivial, then why embark on a multi-million-dollar international project to study it at all? In what way did this reinforce that we were all the same underneath?

For the geneticist and critic Mark Jobling, the way the project was structured, deliberately going after the DNA of isolated populations rather than scanning people all over the world wherever they happened to be, was the thing that ultimately undermined it. "How you define the population in the first place, these are culturally loaded things in themselves. So there was a lot of cultural discrimination in the original aims." The isolated communities that scientists such as Cavalli-Sforza believed were unique were actually never all that isolated nor unique, but they were treated as though they were.

When I raise this issue with him, Greely admits that there was "this sort of uneasy recognition that if your project is about looking for differences then it's sort of counterintuitive to say it's showing similarities." As antiracist as the scientists behind the project were, they had somehow fallen into the trap of treating groups of people as special and distinct, in the same way that racists do. They were still forcing humans into groups, even if they weren't calling these groups races. They were using similar intellectual frameworks to pre-war race scientists, but with fresh terminology.

That's not to suggest that Cavalli-Sforza or his colleagues were being duplicitous. "I don't think he believed himself to be engaged in a racist enterprise," says Joanna Radin. "But I also don't think he really had any more sophisticated a sense of how this was going to fight racism than just being able to show we're all connected, we're all cousins or something." In truth, they saw human differences as meaningful, but they didn't want to focus on that uncomfortable fact in a world in which saying so could have political consequences. Antiracism seemed to be more of a political ideal tagged on as an afterthought. Mark Jobling adds, "They did have a slightly happy-clappy narrative to it, you know, 'joining up the whole human family' kind of thing."

It would have been perfectly possible to study human variation without grouping people. As Jobling explains, the divisions between us are so blurry that humans can theoretically be grouped any way you like. "You could do a thought experiment where you just said we will take Kenyans, Swedes, and Japanese, and will just proportion everybody into those three things." If this were done, because we are all genetically connected to the average Kenyan, Swede, or Japanese person, either directly or by historic migration, then everyone on earth could theoretically be fully assigned to a group based on just these three nationalities. You could say that you were so many percent Kenyan, so many percent Swedish, and so many percent Japanese." This may seem meaningless, but actually it is no more meaningless than dividing the world into black, brown, yellow, red, and white. "The definitions of those populations are cultural, and the choice of population is driven by expediency."

Other geneticists have also warned against dividing up the world. It imposes a certain order onto our species and ignores the actual fuzziness. If the Human Genome Diversity Project had proposed sampling people more systematically, in a grid pattern across the globe perhaps, the true overlapping nature of human variation would be easier to see. Scientists would have been able to map gradual, continuous variation across regions, rather than tight knots centered on very small communities. It's hard not to imagine that this approach—which was mooted at the time but then discarded—might also have been a more effective way of fighting racism. But it wasn't the one the researchers chose.

In the end, most governments, including in the United States, were unwilling to invest in the Human Genome Diversity Project. It never quite got off the ground in the way it was envisioned. And to this day, it remains something of a cautionary tale. In hindsight, part of the problem was that scientists, however well intentioned they were, failed to connect what they were doing with people's real-life experience of race, with the history and politics of this deadly idea. They thought they were above it all, when in fact they were always central to it.

| | | | | | | | | |

Luigi Luca Cavalli-Sforza died in 2018 at the age of ninety-six. When I emailed him shortly before his death, he was retired and living in Milan. I found someone whose commitment to his science hadn't waned, and neither had his personal politics. "There are simply no races in humankind," he wrote back. He was still a hero to biologists in his field, an inspiration, someone who had helped build his scientific discipline into one that today has enormous importance. It was impossible not to admire him.

And, yet, it was also difficult to read his work and come away convinced that his generation of scientists had fully abandoned race science after World War II. Although they had ditched race in name, it wasn't clear that they had necessarily shed it in practice.

In 2000, after the controversy around the Human Genome Diversity Project had been playing out for almost a decade, the project no closer to being realized, Cavalli-Sforza published a book titled *Genes, Peoples. and Languages*. In it, he deftly sketched his grand plan for how genetics could be used to reconstruct human history. It's also a story, the book's back-cover blurb added, that claimed to reveal "the sheer unscientific absurdity of racism."

In a section titled "Why Classify Things?" Cavalli-Sforza wrote eloquently about taxonomy, and how humans have always felt compelled to categorize objects. Yet somehow, he managed to do this without any reference to politics or social history. He never mentioned that humans were classified in large part because it was politically and economically useful to those who did it. He completely glossed over colonialism and slavery, and the ways in which they fundamentally shaped how European scientists thought about race in the nineteenth and early twentieth centuries (indeed, within his own lifetime). Instead, as he saw it, racism was just a scientific idea that turned out to be incorrect. "It seems wise to me," his chapter concluded, "to abandon any attempt at racial classification along the traditional lines."

It's easy to miss the catch. And it lay in the final four words of his statement: "*along the traditional lines*."

"A race is a group of individuals that we can recognize as biologically different from others," he said later. Clearly, then, he hadn't abandoned the use of the word "race" at all. Going by this statement, there may be no room for three, four, or five old-fashioned racial types, with hard divisions between them—the way we usually think about race—but there could certainly be thousands of "social groups" all over the globe, each characterized by their gene frequencies, and therefore having some biological distinctiveness to them. This is the obvious result of relatedness. People who are related are of course closer to each other genetically, and historically we have tended to live near our relatives, which is how clustering of genetic similarity happens. This means that even neighboring towns may be genetically different from one another in some statistically significant way. The small clusters produced by the fact that we don't mate completely randomly could, by Cavalli-Sfroza's definition, be considered "races." So according to his definition, there *are* races, except the number of them is practically endless.

This is an old idea, which owes itself to early geneticists who wanted to move the study of race away from vague generalizations to something more precise. As early as the 1930s, when the field of population genetics was just emerging, Theodosius Dobzhansky—an evolutionary biologist who was later an inspiration to Cavalli-Sforza—was the one to substitute the old-fashioned idea that races were fixed types with the more modern idea that they were populations sharing certain gene frequencies. Like Cavalli-Sforza, Dobzhansky was an outspoken antiracist. But while being actively involved in antiracist efforts within the scientific community, Dobzhansky retained the concept of race. He just redefined it. The way it was redefined squared the circle of how it was possible for all humans to be practically the same while also being different. Under this definition, there's no contradiction in my having possibly more in common genetically with my white neighbor than with my Indian one. My population group as a whole (say, North Indians) will share genes in frequencies that her population group (say, white Britons) doesn't. In other words, if you want it to, race can exist, but you must remember that it's a statistical phenomenon. Not every individual will fit.

"They basically redistributed race," argues Joanna Radin. Race wasn't ditched at all, just the parameters were changed. According to Radin, the problem with the new statistical "population" approach to studying human difference is that even though it may look different in some ways, it hasn't fully shed the baggage of the past. "An interesting analogy would be colonialism," she tells me. "A colonial nation declares independence and they have to forge a new nation with the structures of the old colonial regime, and it's very, very hard to transcend that." Even if the word "race" isn't being used, the idea of race is still there, buried in the bedrock.

The Canadian philosopher Lisa Gannett has similarly warned about the ethical limits of thinking about race in this new way. To some it may not seem racist to think about statistically average "populations" rather than distinct "types" of people. Certainly early population geneticists such as Dobzhansky believed that racism was rooted in the assumption that within one ethnic group people are all the same, whereas those like him believed that, within these groups, people are actually very different. But in the racist mind, as Gannett explains, it doesn't necessarily matter how differences are distributed, so long as they are there in some form or another. This conceptual loophole in population genetics—the fact that we're all different as individuals but that there is also some apparent order to this diversity—is what has since been seized upon by people with racist agendas. Gannett calls it "statistical racism."

The question all this raises is a slightly odd one: Is race still a problem if we redefine race? And even odder: Can science be racist if the people doing it are antiracists? For Radin, intention does matter, but it doesn't fix the underlying problem. "If you look at the UNESCO statements on race, people often think that they declared that race is a social construct and that race doesn't exist. But really what they did is try to constrict use of the term "race" to biologists who could be seen to use it responsibly, and not equate it with inequality." The concept of race was thought to be safe in the hands of liberal, left-wing, antiracist population geneticists because their politics were beyond reproach. "I do think that their sense of virtue really emboldened them to feel like they were creating a transcendent mode of science, that might be able to leave race behind," she explains.

In *Genes, Peoples, and Languages*, Cavalli-Sforza added a humorous aside when talking about the fact that researchers looking hard enough could spot average genetic distinctions between neighboring populations, even at the village level: "People in Pisa and Florence might be pleased that science had validated their ancient mutual distrust by demonstrating their genetic differences," he wrote jokingly. But then, isn't this exactly what racism is? A dislike of other groups in the belief that they are biologically different? In the mind of the racist, it probably doesn't matter how big the groups happen to be, or if the differences are gradual or sharp. It presumably means equally little if it's all about gene frequencies or population averages, so long as the differences are real. If the people of Pisa and Florence could have their mutual distrust validated by population genetics, then why not the people of any other two places?

For Radin, the problem is obvious. It lies in the need to group in the first place, to separate even when that separation means having to zoom in on the very tiniest bits of the genome that might differ, and even then only on average. This need to separate, to treat people as different, is how race was invented. "What happens is that you've got a large community of very well meaning, self-described antiracist scientists seeking to find a way to move beyond race into population genetics, which seems to be incredibly neutral. It's numbers, it's statistical, it's objective," she says. "What they have a more difficult time reckoning with is that even something like population genetics is a science done by people, working with the assumptions and the ideas that are available at the time." They may believe themselves to be free of racism, but they can't help thinking about humans in racial terms.

In his correspondence with me, too, Cavalli-Sforza made one comment that surprised me. He observed that interracial relationships—relationships that, in the early twentieth century, eugenicists feared might lead to offspring with strange physical and mental deformities—has turned out to be no bad thing. Miscegenation, as it was called, is now obviously recognized to be no threat to human health. You only have to look at "the beauty and vitality of hybrids, children of partners coming from genetically distant groups," he wrote in his message. To use the phrases "hybrid" and "genetically distant" might have seemed at one time scientifically acceptable, but is it anymore?

To make this observation and use this language in the twenty-first century just feels plain odd.

There is a gray area in which well-meaning people make what in other contexts might be considered incendiary statements, and we overlook them because we know they are well-meaning. In reference to reproduction rates, Cavalli-Sforza's book stated, "Europeans are largely at a standstill while populations in many developing countries are exploding; thus blonds and light-skinned people will decline in relative frequency." If the superficial differences between us don't matter, then why should this? To the population geneticist, that people with blond hair are disappearing may be as much of a concern as the possibility that Native Americans might dwindle, or that indigenous Andaman Islanders living in an archipelago in the Bay of Bengal might be subsumed into the wider Indian population. To the antiracist, objective scientist, there is no value judgement in this. It's a problem to science only because researchers are losing some interesting subjects of study, some statistical corners of human diversity, perhaps a few rare blue-eyed-blond gene combinations. But to someone with alternative politics, this might be seen instead as an argument in favor of racial purity, of preserving distinct population groups against the threat of miscegenation.

It's easy for academics to imagine that the language they use, and the frameworks they operate in, don't really matter. They are just words, not data. "I think that in the real world what the scientists say has about as much influence as turning on a fan has on El Niño. It's what throwing a cobble into the English Channel has on Atlantic weather," Henry Greely tells me near the end of our interview. "We're just not that important." But it does matter, because their frameworks and language contribute to our understanding of ourselves. If scientists call people of mixed ancestry "hybrids," this implies that race is real because we are different enough to warrant using that word. If they talk about "isolates," this sounds like there are groups who are more "racially pure" than others. Dismantling the edifice of race is about more than just tweaking language; it is about fundamentally rewriting the way we think about human difference, to resist the urge to group people at all.

It takes some mental acrobatics to be an intellectual racist in light of the scientific information we have today, but those who want to do it, will. Racists will find validation wherever they can, even if it means working a little harder than usual. And this is the reason that good scientists who do reliable research, ones who are also well-intentioned and antiracist, like Cavalli-Sforza, can't afford to be cavalier or leave too much room for misinterpretation. There's an uncertain space between recognizing that there is a gap of knowledge and actually filling that gap. It's a place where speculation thrives, where the racists reside. Racists adopted the same concepts as good scientists—and the same language as antiracists—to assert that, if certain groups can show some average differences to other groups then, by that logic, certain groups *might* be better on average than others at certain things. When Steve Sailer and his followers talk about "human biodiversity," this is what they mean. This wolf in sheep's clothing is twenty-first-century scientific racism.

And we were told this might happen. A caustic report on the Human Genome Diversity Project released in 1995 by UNESCO's International Bioethics Committee sounded precisely this warning. It argued that the project, whether it meant to or not, could give racists some basis to believe that certain groups were inferior or superior to others. In particular, the committee was concerned that by bringing genetics to the fore in telling the human story, people would ignore culture and history, and return to the kind of simplistic biological thinking that propelled the eugenics movement in the early twentieth century. It advised scientists to resist the temptation to use their work to shore up any kind of political ideology, whether racist or antiracist. "Racism," it reminded them in case they had forgotten, is "socially and politically constructed."

Science is just a pawn in the bloody game.

| | | | | | | | | |

Although it never got off the ground, the concept behind the doomed Human Genome Diversity Project did survive. In the years that followed, other teams stepped in to achieve essentially the same outcome in other ways. The Center for the Study of Human Polymorphisms in Paris today keeps a bank of DNA samples from populations all over the world, ready for researchers who want to tap it. In 2002 the United States National Human Genome Research Institute introduced a $100 million initiative to study human variation. And in 2015 the United Kingdom launched its own project to make a genetic map of the people within its own borders, named People of the British Isles.

The project had one more unintended consequence. In 2005 the National Geographic Society in Washington, DC, the one behind the famous magazine and satellite TV channel, decided to dip a toe into the world of population genetics. Naming its effort the Genographic Project, it chose Spencer Wells, an anthropologist, geneticist, and television presenter, to lead the project. Wells had spent a portion of the 1990s studying under Luigi Luca Cavalli-Sforza, seeing the difficulties around the Human Genome Diversity Project at close quarters. His solution to the controversy was simple. National Geographic would sell easy-to-use kits that members of the public could use to help them understand the history of migration that might be hidden in their DNA, and in the process the organization would build a data bank from their genetic information.

"We put together this consortium of scientists with the goal of sampling the world's DNA, and at the same time wanted to enable anybody, any member of the general public who was curious about their own genetic ancestry, to get themselves tested," Wells tells me. The idea that people might want to spit into a cup and have their ancestry tested didn't seem at the time like a highly profitable venture. Even the CEO of National Geographic was doubtful, warning Wells before the launch that nobody was going to spend a hundred bucks to test their DNA.

The project turned out to be a money maker. "The day we announced, we sold ten thousand. It had hit a hundred thousand by the end of the year," says Wells. "It launched the consumer genomics industry"—which got a noticeable boost in 2006 when the media legend Oprah Winfrey had her DNA tested for a television show, revealing ancestral links to people now living in Liberia, Cameroon, and Zambia. It also turned out that Winfrey, unlike many black Americans, shared no recent ancestry at all with Europeans. "I feel more connected to where I've come from," she told the show's host, Henry Louis Gates Jr., a Harvard professor of African American studies, who wrote a book about her experience.

Before long, companies such as 23andMe and AncestryDNA were selling their own kits, turning over billions of dollars. In 2018 it was announced that AncestryDNA alone had sold a total of around ten million kits around the world.

Spencer Wells has since left the Genographic Project, become the owner of a nightclub in Texas, and moved on to new genetic testing ventures. He tells me that the personal ancestry testing industry flourishes in the United States because of the rootlessness of so many of its citizens. "In societies like the present-day US, where we have a lot of hyphenated Americans—Irish Americans, Italian Americans, Hispanic Americans—people feel somewhat disconnected from the entity that comes before the hyphen. So they want to figure out who those people were. Who were the people who migrated to the US?" For black Americans in particular, most of whose ancestors were transported as slaves and ripped away from any connection to their families or homelands, the kits have offered the only means they may have of tracing their genealogy. But the psychological effect on the public of sequencing the human genome, of convincing people that our differences are identifiable in our DNA, is that this now appears to be a reliable way to define who we are.

In reality, genetic testing is only an educated guess about where your relatives may have lived, based on the data fed into the models in the first place. As geneticist Mark Jobling explained, it's possible to group people any way we like. Ancestry tests scan portions of people's genomes to find those who have genetically a little more in common, then pool them together. Theoretically, they could pool them using any measure. But, of course, companies most often use old-fashioned racial categories or nationality.

This also means that if there are no DNA samples from the modern-day people related to you, you're stuck. For example, one of the reasons Winfrey is linked to Liberia may be that this is where former slaves were long ago sent by white American leaders who couldn't bear the thought of slaves living freely among them. Ancestry testing doesn't show you your past as much as it reveals the people you are distantly related to in the present who have had similar tests done. Oprah has some connection to people who now live in Liberia, but this is not necessarily her ancestral homeland, from where her relatives originally came.

Mark Thomas, a leading geneticist at University College London who has seen his own research recruited into these models, tells me he has always been skeptical of ancestry testing firms. "It's not that they're cynical, it's not that they're nasty, it's not that they've got particularly racist agendas. They want to make money, and you make money by servicing peoples prejudices." Not just prejudices, but also people's desire to know *who they are*. Thomas and his colleagues were threatened with legal action by BritainsDNA, one such company based in Scotland, after they challenged the firm's wide-ranging claims in the press, for instance, that the actor Tom Conti is "Saracen" in origin, and that Prince William has Indian ancestry. Neither claim could be established by genetic data.

Ancestry testing has taken the work of well-meaning scientists who only tried to do good in the world and inadvertently has helped reinforce the idea that race is real. Using the methods and data of scientists such as Cavalli-Sforza, an entire industry has achieved exactly the opposite of what these scientists once set out to do. The true way that variation works, the nuances, are rarely explained by those selling ancestry-testing kits. Having seen how "black" Oprah really was, white supremacists in the United States began using the very same tests to prove how "white" they were, sometimes sending off vials of spit to various companies until one came back with the desired result and established beyond doubt that they were of purely European ancestry. By tracking history through our genes, by dividing up individual bodies into proportions of nationalities—so much European, so much African, so much Asian—the tests fortify the assumption that race is biologically meaningful. If it's possible to categorize, we assume, there must be something to the categories.

The irony is that as more research has been done into our origins since the launch of the Human Genome Project and consumer ancestry testing, it has only undermined these measures of identity. Within the last decade, as scientists have uncovered exponentially more genetic evidence about us and our ancestors, even they have been surprised by the results. Nothing has matched expectation. Our roots, it turns out, are very rarely where we think they are.

CHAPTER 7

# Roots

## *What race means now in the light of new scientific research*

When she was growing up, my little sister was a diehard fan of Morrissey, the frontman for the Smiths, hailed as a genius songwriter and British cultural icon. For one of a handful of brown girls in a white working-class Southeast London suburb, indie music spoke to that cold, lonely feeling of not quite being able to fit in. If the far-right British National Party was marching outside our door, its members calling for an all-white nation with no immigrants, inside her headphones was a different British voice that she could relate to. He was a refuge from those who insisted that we all had to be the same.

But in an interview with a music magazine in 2007, Morrissey said something that couldn't help but trouble my sister, as well as other fans. "Whatever England is now, it's not what it was and it's lamentable that we've lost so much," he complained. He railed against high immigration, against what he saw as a change in the character of Britain. There was public outrage. She lost a hero. But as we in our family knew too well, out in the country as a whole there were many who felt this way. This was a debate that had been simmering for decades, occasionally stoked by national politics, making people anxious, wondering what it meant to be British.

A decade later, the pot bubbled over. A financial crisis and economic austerity, coupled with higher-than-usual rates of immigration from eastern Europe, helped fuel support for nationalists who wanted to cut the country free from the European continent. In a referendum in 2016, the majority of voters agreed that leaving the European Union might be a good plan. They were promised a new dawn. The nation would stand alone, the way it had done during the days of empire, riding the waves of unbridled trade and setting its own rules on who would be allowed into the country.

For visible immigrants, or children of immigrants like my sisters and me, watching this play out could sometimes feel like an out-of-body experience. The borough in which my parents lived and where we grew up was one of only five out of the thirty-two in London that voted to leave. As citizens we had the right to vote to decide Britain's future, but we also knew that a sizable slice of other voters wanted fewer of us there in the first place. A campaign poster showed legions of men with skin as brown as ours queuing up in front of the slogan "Breaking Point." The far right was emboldened. Around the time of the referendum, reports of race-based crime rose, and there was a sharp spike in the kind of everyday racism that I had last seen as a teenager.

Squaring your appearance with your nationality is one of the hardest parts about being a member of an ethnic minority. Not all, but some of those who voted to leave Europe wanted a return to their own particular vision of Britain. Skin color mattered to them because skin color—white skin—was a visible baseline, a reference point. White was how the British had always looked, from the beginning, before empire, before Shakespeare, before kings and queens, before culture and values. Britain, as far as we were aware, has been forever white. In their eyes my failure to be the right color truly undercut my claim to Britishness.

Nobody could have predicted then that, by an almost cosmic coincidence, at the very moment Britons were struggling to define their identity in the face of political turmoil, and particularly for those racists who saw Britain as a white nation first and foremost, some news was coming. They were about to be thrown a curve ball.

| | | | | | | | | |

I saw it for myself at London's Natural History Museum in early 2018, a package no fancier than a bunch of old bones.

The skeleton is laid out neatly in a small corner of the museum. Most of the visitors don't linger as I do. To be honest, it looks unremarkable. But this is the frame of one of the oldest dead bodies ever found in the country, some ten thousand years old. And it's full of secrets. Almost as soon as the bones were discovered in caves in Cheddar Gorge in Somerset in 1903, giving their owner the name Cheddar Man, people began to wonder how this individual must have looked. They wanted to put a face to one of our early ancestors. Archaeologists could certainly guess that he was slightly short by modern standards, that he probably had a good diet, and that he may have been around twenty years old when he died. One speculative reconstruction showed him to be white-skinned, with rosy cheeks and a trailing brown mustache. But his actual appearance was a mystery.

This is where the genetic study of the bones of our distant ancestors, of their ancient DNA, came in. It has achieved what the Human Genome Diversity Project couldn't. The DNA of living people offers a limited and fuzzy picture of the past. When it comes to tracking human migration patterns over thousands of years, even archaeology and linguistics can't provide all the detail that ancient DNA can. By around 2010 genetic sequencing techniques had developed far enough to tease out highly reliable samples of DNA from ancient specimens (a bone just behind the ear turned out to be best) and use them to help reconstruct entire genomes of long-dead people. The use of this technique has mushroomed in the last decade. It has been credited with solving historical mysteries at a stroke. Thousands of skeletons from all over the world have been analyzed already, and as the British public were about to learn in early 2018, Cheddar Man was one of them.

Scientists at the Natural History Museum and University College London revealed that Cheddar Man probably had blue eyes and curly hair—no great surprise here. But what came as a real shock to many Britons was that his bones also carried genetic signatures of skin pigmentation more commonly found in sub-Saharan Africa. It was probable that Cheddar Man had dark skin. So dark, in fact, that by today's standards he would be considered black. The revelation, along with a dramatic new reconstruction of his face markedly different from the original one, made front-page news and television bulletins:

> "Hard Cheese for the Racist Morons," ran a headline in the tabloid *Mirror.*
>
> "Another Racial Panic for White Supremacists," announced the news website *Salon.*

Panic was indeed sparked. There were all the stages of grief. On far-right websites, a few immediately began doubting the scientific results—maybe, just maybe, the researchers had gotten it wrong. Some hopefully voiced the possibility that Cheddar Man hadn't been an actual Briton at all, but was just a passing visitor who happened to die here, like an unlucky tourist. Finally, there was acceptance. Some, especially those who for so long had believed that skin color was the basic measure of Britishness, wondered if perhaps it was time to rethink national identity.

If the original Britons were black, all bets were off.

Throughout the frenzy, there was one set of people for whom the news barely registered a flicker on their excitement dial. They weren't shocked at all.

"With the whole Cheddar Man thing, I was amazed initially at just how much press coverage it got," I'm told by Mark Thomas, of University College London, who worked on the finding. Leaning back in his chair casually, wearing stonewashed jeans and a granddad-collar shirt, Thomas is about as relatable as professors come. He is one of the world's leading experts on ancient DNA, and from this position of authority he has a tendency to tell it how it is. For geneticists like him, the Cheddar Man discovery was unremarkable given what they already knew. They had more or less expected it.

Thomas had welcomed the discovery as just another piece of evidence in a huge body of research. It was a couple of sentences in his latest paper. Scientists had already known for a few years, from analyzing the skeletons of other hunter-gatherer bones found in western Europe, that dark skin pigmentation could well have been common back then. After all, light skin was likely an evolutionary adaptation, one that helped people living in northern climates absorb more vitamin D because there wasn't enough sunshine. The first human pioneers probably didn't arrive in Europe or Asia looking white, because they had originally migrated from Africa, where there was little or no survival advantage in having little skin pigmentation and hence, lighter skin.

What researchers were a little less sure about was how quickly lighter skin emerged, where and when. "Over the last ten thousand years? Or over the last forty thousand years?" asks Thomas. One theory is that it developed very slowly and gradually over the last roughly forty thousand years, since modern humans started to live in Europe. Another theory is that it was a more recent phenomenon, which came perhaps with the advent of farming. Trading a hunter-gatherer lifestyle for settled agriculture would have limited people's diets and made it even more vital that they get the vitamin D they needed from the action of sunlight on their skin rather than from their food. Another theory is that light skin emerged elsewhere in the world, outside western Europe, and that the movement of peoples would have then introduced it into the darker-skinned European populations. Evidence as it stands indicates that, like the Cheddar Man, many other pre-farming hunter-gatherers who lived in western Europe during this time and at least up until as recently as seven thousand years ago would have had light eyes, dark hair, and dark skin, and that the first farmers to come into the region later from the east brought with them their lighter skin and brown eyes.

One thing was clear: Cheddar Man wasn't an exception in his time. People all over the world then didn't look anything like the way we look now. Not only this, they looked more different from each other than we do today.

"Differentiation between groups in different parts of the world would have been greater," explains Thomas. The scientific explanation for this is genetic drift. Being in small groups as they were, every breakaway bunch of migrants as it moved began to look more and more different from the relatives they left behind as time passed. Since then, as groups grew bigger and remixed with each other, populations across the world have become more homogenized. Ten millennia ago, we would have struggled to identify a person's geographical origins by modern measures of race and ethnicity. Appearance didn't map the way it does now, and physical features may have been dramatically different in different regions.

When Thomas and his team studied the very earliest farmers in the Fertile Crescent, who lived in what is now Iran, and compared them with farmers in nearby Anatolia and the Aegean, they found to their surprise that the two were genetically very distinct from each other. "They were as different as people from Ireland and Thailand today, more or less. I mean, of that order of magnitude." Today, neighboring populations tend to be much more similar. They've mingled and mated with each other, mostly dissolving away the gaps.

Yet, our modern ideas of race are deeply connected to how we look. Our appearance is a shorthand for the stereotypes, a means of slotting people into groups and making judgments about them. The disbelief that met Cheddar Man's probable blackness was because many among the British public couldn't help but assume that Britons had always looked a certain way, even in the distant past. They struggled to categorize Cheddar Man, forgetting that he existed thousands of years before our racial categories came about. He was proof that there couldn't be anything eternal or pure about race because once upon a time, not so very long ago in evolutionary terms, most of the people on earth didn't look like us. They were already human. They were, however distantly, *us*. But they looked different.

The picture becomes even more complex as we go further back in time. Sarah Tishkoff, a geneticist at the University of Pennsylvania who has led pioneering studies into skin-color variation across Africa, has found that the genetic variants—different forms of the same genes—associated with both dark and light skin have existed in Africa for a long time. The variants that are associated with light skin were common not just in Europe and sometimes in East Asia but also among the San hunter-gatherers. "These are the people in southern Africa who have the oldest genetic lineages in the world," she says. This suggests that rather than occurring independently outside of Africa as new mutations, gene variants associated with light skin may already have been part of the human genome when people first migrated out of the continent.

Thus, not only did people with darker-pigmented skin occupy Europe, but even earlier, there were genetic variants for lighter-pigmented skin in Africa. Given the evidence so far, Tishkoff suggests that lighter-pigmented skin may even have been the ancestral state in the long distant past. Underneath their bodies, chimpanzees, our closest genetic relatives, tend to be light skinned, their dark body hair providing protection against the sun. "When our ancestors left the forest and went to the savannah, there would have been selection for better thermoregulation, so getting rid of body hair, increasing the number of sweat glands. And if you're decreasing the body hair, there would be selection for darker skin." Darker skin could have been one of the adaptations to a new living environment within Africa.

When scientific reconstructions are made of earlier human species, such as *Homo erectus*, they are almost always given dark skin. However, says Tishkoff, "I don't think that's necessarily the case, because both light and dark variants have been around for really long time. And there could have been variation in Africa a million, two million years ago." The long lens of evolutionary history has a way of turning all you think you know on its head.

Even today, there is far more variation in Africa than the simplistic black-white model of race implies. "I think many people don't recognize the large range in skin color in Africa," says Tishkoff. "The whole continent of sub-Saharan Africa is incredibly diverse genetically. It doesn't fit with a racial model, one homogeneous African race. There's a huge amount of variation amongst populations in Africa. Skin color is a terrible racial classifier. There really are no good biological classifiers for race."

For the biologists who know this, skin color begins to lose its meaning. "I mean, it's skin pigmentation, you know! It's just so trivial," says Mark Thomas, which is why he found reactions to the new finding about the

Cheddar Man bizarre. "Obviously there are some idiot racists over there in the corner for which it is important. But I think that if you base your identity on the pigmentation of some West Country bloke from ten thousand years ago then you really should rethink it. My own personal view is that today we over-privilege and fetishize the concept of identity."

Thomas reminds me that the physical features we associate with race are poor proxies for overall genetic similarity. Even if one population tends to have darker skin and another lighter, that doesn't mean their genomes will have less in common than two populations with the same skin color. Variations in physical appearance, whether it be skin pigmentation, ear shape, nose shape, whatever, says Thomas, make the gaps between groups feel far larger than they really are, genetically. Biologically, the differences really are just skin-deep. It's an error to assume that the internal differences are as profound as the external ones appear. But it's an easy one to make. "If we could see each other by looking at our genomes, you would be hard pushed to work out whether somebody was from India or from Poland. You'd be hard pushed on the same number of variants," he explains. "There is relatively little genetic differentiation between Southern India and Ireland. I mean, relatively similar ancestry components. But of course, the pigmentation differences are quite large, and so people assume that these people are massively different genetically." In that sense, how we look is misleading. "Nature plays dirty tricks on us," says Thomas.

It can play tricks on scientists, too. If data seem to suggest that populations are very different, it's largely because population geneticists are deliberately examining the small sections of our largely shared DNA that happen to differ. This is their job. "We're zooming in. We're turning up the contrast on what are actually tiny little differences over extremely closely related populations," Thomas warns.

| | | | | | | | |

"The past is very surprising," David Reich, a geneticist in the ancient DNA laboratory at Harvard University, tells me. "It's different from how most people picture the past in their heads."

Reich is the most well-known person in this branch of science, at the forefront of the science of using genetics to plot ancient migrations around the world. At the moment I happen to visit him, though, he has become embroiled in controversy for suggesting in the press that more work needs to be done to understand cognitive and psychological differences between "population groups," a phrase that most people have interpreted as meaning "racial difference." His statement—a departure from the nearly seventy-year consensus that studying race isn't the business of mainstream science—has attracted angry emails from fellow academics. But he hasn't backed down. When I see him, I expect him to be defensive, maybe even brash.

I couldn't be more wrong. With his hands in his lap, so soft-spoken that my voice recorder barely picks up every word, he surprises me with his gentleness. His half of his office is bare, save for a few drawings stuck to the plain white walls. He is unfailingly polite, pausing only to message his wife. The one clue to his global importance as an academic is the steady stream of students and researchers lining up to see him outside. One young man sits with his laptop at a bench in the corridor all day in the hope Reich may be able to spare him a minute or two later on.

Reich's lab is a powerhouse. It has scoured the world for skeletons that might provide genetic evidence of the past, and as Reich has noted, it churns out findings so quickly that the amount of data doubles faster than the time it takes to publish new data. Scientific journals simply can't keep up. But for him, this is more than a scientific gold rush. Genetics has a way of cutting through ancient historical questions in a way that nothing else can. His group, along with the lab of Mark Thomas and others across the world, helped confirm a longtime hypothesis: that farming emerged ten thousand years ago in the Near East—the region between Europe, Africa, and Asia—among genetically varied groups of humans who then helped spread agriculture to other regions. He is also fairly confident that natural selection has caused southern Europeans to be a little shorter on average than northern Europeans—this must have benefited them in some way.

But it's the story of migration that is the most revealing. What we think of as "indigenous" Europeans are, Reich and other scientists now understand, the product of a number of migrations over the past fifteen thousand years, including from what is now called the Middle East.

The British have their own story to tell. "Britons in the past didn't look like Britons today, and were genetically very unlike Britons today," Reich says. Whoever the first inhabitants of Britain were, their way of life was likely to have been almost totally replaced around 3,000 to 2,000 BCE by a group of people who traveled through Europe from the steppe grassland that stretches between the Black and the Caspian Seas. They are known by some anthropologists as Beaker folk for their distinctive bell-shaped pottery. Artifacts of Beaker culture are found scattered all over Europe; now, the Harvard team has shown by studying the DNA of four hundred ancient Europeans that these people must have swept in and supplanted almost everyone who was living in Britain at the time as well.

How they did this is unclear. They could have simply come in large numbers and bred with people who were already there. They may have been better equipped to survive in the environment, through resistance to certain diseases or by virtue of their technology. The preexisting populations could already have been collapsing, as some data suggests. Whatever the explanation, their arrival changed not only the culture but also the way people looked. The steppe people with their Beaker culture also had lighter skin. According to estimates drawn up recently by Reich and his colleagues, this Beaker invasion replaced around 90 percent of Britain's gene pool in the space of just a few centuries.

This means that light skin did not define Britons from the beginning. "There's been a continuous process of skin lightening, with big jumps that occurred at these migrations," Reich explains. "So, for example, when the first farmers came to Britain about six thousand years ago there is a big change in the average hue of skin around then, predicted by the genetics. And then when these Beaker people spread into Britain, another big jump was associated with that."

Some of this work confirms what archaeologists already suspected, but what has been really surprising is just how much churn there has been in global patterns of migration throughout the ages. Reich himself was taught when he was younger that humans spread out of Africa, with little mixing once they started to split off, like branches of a tree. Once they landed somewhere, people stayed put. That was the common assumption. But the evidence that's now emerging suggests something entirely different. "It became very clear that the big large-scale mixture, migration, or gene flow, however you want to call it, is common and recurrent."

The true human story, then, appears to be not of pure races rooted in one place for tens of thousands of years, but of ongoing mixing, with migration constantly changing geographical direction. The cherished belief that people in certain places have looked the same way for millennia has had to give way to the understanding that migration made the world a melting pot long before the last few centuries, long before the multicultural societies we have today. Our roots are not like an orderly family tree but instead are tangled, according to Reich, more like a climbing plant on a trellis. Our ancestors branched out but then came back, and remixed, again and again throughout the past.

"I think this idea of indigeneity, and you being from a population that has been here for ages—I mean there may be populations that have better claims to that than others—but at some deep level the great majority of people in the world, if not everyone, is not derived directly from people who lived in the same place deep in the past," he says.

The British story is just one of thousands. For example, the Beaker folk were part of a far earlier, bigger, and longer migration out of central Eurasia and into many different corners of the world of people associated with what archaeologists call the Yamnaya culture. They were pastoralists, raising and moving livestock, with wagons and horses that made them mobile in a way that may never have been seen before. Their diet was rich in meat and dairy. From roughly seven thousand years ago to roughly five thousand years ago, the Yamnaya (themselves a product of earlier migrations into the region they came from) trekked west and southeast, populating not just Europe but also as far as northern India. They introduced the wheel and, it has been suggested, also cannabis to the regions they migrated to.

By 3000 BCE, the Neolithic (New Stone Age) farming cultures of Europe had been pretty much replaced. Kumarasamy Thangaraj, at the Centre for Cellular and Molecular Biology in Hyderabad, tells me that around the same time, people of the Yamnaya culture came in from the north of India and mixed with the people who were already there. The Indian population they met was itself a mix of indigenous hunter-gatherers who had originally moved out of Africa many thousands of years earlier and more recent farmers who had migrated from what is now Iran. All Indians, save a tiny community of Andaman Islanders who have been isolated from the Indian mainland for thousands of years, are a blend of these three ancestral populations.

Confirmed by gene studies, these ancient connections can also be spotted in the words we use. Linguists long ago saw remarkable similarities between European and Indian tongues, describing them together as Indo-European languages. Genetics has added more hard data to the history. Almost all Indians today are genetically closely connected to Europeans by their ancient ancestors who spread the Yamnaya culture, as well as the earlier spread of farmers from the Middle East.

| | | | | | | | |

"If you pay any attention to the discoveries coming out of science, they don't play into any sort of old systems of prejudice," David Reich tells me.

Take Stonehenge, the mysterious prehistoric assembly of standing stones in southwestern England, which attracts more than a million visitors every year. Within a few hundred years of its construction, around five thousand years ago, the Neolithic farmers who had built it were pretty much gone. They were probably replaced by incoming folk who followed the Beaker culture, because Reich's team couldn't see much evidence of Neolithic ancestry in the genomes of ancient human remains they were studying that dated from four thousand years ago. Now, just pause to think about what this means: the symbol people associate with ancient Britain, the one thing that couldn't really be more authentically British, was built by people who are certainly not the main ancestors of those who consider themselves indigenous Britons today.

Cheddar Man and his relatives, too, who lived ten thousand years ago, couldn't have been from the same genetic pool as Britons today, because like the builders of Stonehenge, they were replaced by farmers who had spread across Europe from Anatolia. "Cheddar Man and his people formed a unique genetic cluster. They don't have any direct descendants—only bits of them exist," explains Mark Thomas. By "bits" he means that Cheddar Man and his relatives who lived on the continent would have bred with whoever came into the region. So although his own particular population and their culture didn't survive intact, traces of them would have endured, either because they mixed with farmers coming into Britain or because their continental relatives mixed with farmers spreading across Europe. According to Thomas, roughly 10 percent of the ancestry of most modern Britons is shared with Cheddar Man's genetic group.

To know that this melting pot has been churning for thousands of years puts a fresh spin on the contemporary idea of race. "I think that genetics and genomics have a wonderful opportunity to undermine these outdated and scientifically unsupported notions of race, ancestry, ethnicity, and identity," says Thomas. The feeling that there is a "home" for us all, and that our bodies somehow reflect this, deeply and viscerally, begins to melt. The attachments that some people have to places and their relics, the ancient stories they construct around who "our people" were, have to be rethought when they learn that "our people" were actually migrants into a place occupied by others. The relics belong to them. Almost everyone on the planet is the descendant of a migrant from somewhere.

What is even more mind-bending is that when you look this far back in time, ancestry expands to include almost everyone. "Cheddar Man's people are to an extent the ancestors of just about everybody in Europe," he explains. "Indeed, it is possible that in his group are the ancestors of everybody in the world, just about; maybe everybody in the world today."

This may seem implausible, but it's just mathematics. The further back you go in time, the weaker your genetic link to your ancestors. Five generations ago, you would have as many as 32 possible ancestors contributing to your genetic makeup. Nine generations back, you could have 512, many of whom may have contributed next to nothing to your genetic makeup. Zoom back fifteen generations—still just the tiniest slice of recent human history—and you could have 32,768, assuming nobody was having babies with someone they were even distantly related to, which is unlikely. Each of these could give you no more than the tiniest fraction of your DNA. Longer and longer ago, the theoretical number rises into the millions—ultimately, to more people than were even alive at the time. Of course that is impossible. The only explanation is that we are all at least a little inbred.

Even if you could trace your lineage as far back as Cheddar Man, or even more recently, to Cleopatra, Julius Caesar, or any other figure from antiquity, you would probably be no more related to that individual than is any other person on the street. The more you zoom into the past, the more your ancestral history begins to overlap with that of everyone else on the planet. As Thomas notes, we only have to go back a few thousand years before we reach somebody who is the ancestor of everybody alive today. Go back a few thousand years more, and everybody who was alive is either the ancestor of everybody alive today (if they had descendants who survived), or nobody alive today (if they didn't). Hence Cheddar Man, if he had children and they had children, and so on until today, is both your relative and mine.

But what does this have to do with national identity? Many would argue that identity isn't necessarily about biology or appearance, it's actually about language, culture, and values. But if it's about language, culture, and values, then I am as British as anyone. I was born in London, I speak the Queen's English, and I live a quintessentially English kind of life. My dinners generally comprise roast meat and vegetables (I probably eat a curry as often as most white Britons, and possibly less than some), my radio is set permanently to the BBC, and I celebrate Christmas alongside almost everyone else. If skin color and genetic purity can't be a measure of ethnic identity, because Britons have changed on both these counts over the millennia, then there's nothing to prevent anyone from anywhere from earning citizenship and becoming truly British, even by the most conservative standards.

When considered from the perspective of the deep past, race, nationality, and ethnicity are not what we imagine them to be. They are ephemeral, real only to the extent that we have made them feel real by living in the cultures we do, with the politics we have. David Reich tells me that he draws a sense of global kinship from his work on genetics. "I have a personal way in which genetics is meaningful to me which doesn't involve my own ancestry," he says quietly. "I think that one way of relating to the findings about genetics is that we're all related to each other, and we are all part of a broadly closely related group of people over the last couple of hundred thousand years, with a lot of complexity, and with a lot of mixtures and migrations and reticulations. And we're all part of that."

But then his tone changes. Even after everything he's said, he doesn't dismiss the idea of race altogether.

| | | | | | | | | |

David Reich isn't a racist. But neither does he adopt the staunch antiracist position of the old-school population geneticists, such as Luigi Luca Cavalli-Sforza, who bravely debated the scientific racists of their time, who wore their politics on their sleeves. Reich respects Cavalli-Sforza, even writing about how much he has been an inspiration to him. But he confesses that he sees himself as apolitical.

The genetics of human variation are complicated and subtle, he tells me. And his own position on race is a similarly subtle one. Despite his research revealing the extent of interconnectedness between humans, the great uniting trellis of ancient migration, Reich still suspects there's something worth investigating about group difference. And he leaves open the possibility that this difference correlates with existing racial categories—categories that many academics would say were socially constructed, and not based in biology at all, except for in very unreliable ways, such as along crude skin-color lines. "There are real ancestry differences across populations that correlate to the social constructions we have," he tells me firmly. "We have to deal with that."

He admits that some categories make no biological sense, such as the way "Latino" is used in the United States to refer to anyone from South America. "'Latinos' is a crazy category that encompasses groups with different ancestry mixes ranging from Puerto Ricans, who have very little Native American ancestry, mostly African, a little European; to Mexicans, who have very little African ancestry and [are] mostly Native American, European. . . . It's a crazy category." At the same time he thinks some categories may have more biological meaning to them. Black Americans are mostly West African in ancestry and white Americans tend to be European, both correlating to genuine population groups that were once separated at least partially for seventy thousand years in human history. "There's a long time separating these two groups," Reich says. "Enough time for evolution to accumulate differences. We don't know very much about what those differences are because we're still at the beginning of collectively trying to identify biologically what differences do."

He suggests that there may be more than superficial average differences between black and white Americans, possibly even cognitive and psychological ones, because before they arrived in the United States, these population groups had this seventy thousand years apart during which they adapted to their own different environments. Reich implies that natural selection may have acted on them differently within this timescale to produce changes that go further than skin deep. He adds, judiciously, that he doesn't think these differences will be large—only a fraction as big as the variation between individuals, just as biologist Richard Lewontin estimated in 1972. But he doesn't expect them to be nonexistent either: as individuals we are so very different from one another that even a fraction of a difference between groups is *something*.

They are words I never expected to hear from a respected mainstream geneticist. I know that Reich is not a racist. Indeed, like Cavalli-Sforza, he believes that if race research is done, it will only further demolish old prejudices. Scientists are concerned with fact, not fiction, his argument goes, and the facts we have accumulated so far are simply not in the racists' favor. The more good work we have, the more it demolishes longstanding racial stereotypes, so there should be no barriers to doing even more research, even if it feels risky. Reich tells me, "My feeling about this field has been that, broadly, it makes telling falsehoods more difficult. That's my feeling. It may be self-serving, but that's my feeling. And so I think these surprises, such as ancient Britons were very much more dark skinned than present people, . . . I think this is broadly a force for combating prejudice, because it doesn't conform to anybody's pictures they had before."

Though Reich sees the racists as factually wrong, he also sees some antiracists—those who insist that we are all exactly the same underneath—as not having the full facts either. "It's a little bit painful to see very well-meaning people saying things that are contradicted by the science, because we want well-meaning people to say things that are correct," he says. "The way I see what's going on in this world right now, there are racist people that are just perpetrating falsehoods, and just representing the science in incorrect ways, tendentious ways in order to achieve certain goals. And then there's people whose perspective on the world I agree with who are actually saying things that are technically incorrect."

Reich is technically correct that there could be more profound genetic differences between population groups than we are aware of at the moment. But to date, no scientific research has been able to show any average genetic differences between population groups that go further than the superficial, such as skin color, or that are linked to hard survival, such as those that prevent a geographically linked disease. There is no variant of any gene that has been found to exist in everyone of one "race" and not in another.

In London, Mark Thomas, who has collaborated with Reich, remains dismissive of the idea that race as a concept is useful to the study of genetics. "Most researchers, including geneticists, agree that 'race' is a socially constructed category. There is no categorical imperative in biology, and no need or value in placing people in biological boxes. There are subtle genetic correlations with geographic origin, and physical traits, as well as medical ones, and understanding those correlations is important. But there are no hard borders, just gentle gradients," he tells me. "Unfortunately, that doesn't stop people 'racializing' others, and perhaps that reflects our desire to categorize. Most categories are nonsense, although some may be useful. 'Race' is useless, pernicious nonsense."

The question of whether or not biological research into racial difference is useful still divides the scientific community. What seems to bother Reich above all is not that deeper racial differences may exist, but that biological research on these differences isn't being done, at least not properly and not enough, so we just don't know. Of course part of the reason for this is the longstanding scientific taboo against what might be considered race research, which has kept race off the table in mainstream biology since the end of World War II—although certainly not off the table in social science, which has built an enormous body of work on the topic. We have plenty of data on racial gaps in income, health, and schooling. The reason for this is that race has been accepted by academics as a social reality, not a biological one. Race affects how we live, but not who we are genetically.

Reich, however, appears to find this unfair. "We've been silenced by the great anxiety that we feel talking about these things, and by the history of abuse of genetics by people seventy years ago or eighty years ago," he says.

He is probably not the only scientist who would like to be free of "the great anxiety" caused by eugenicists and scientific racists in the past. But that freedom would also have to come with responsibility. As the devastating mistakes of the nineteenth and early twentieth centuries proved, race research never goes well when society is racist. Although Reich insists that biological data as it stands makes racism impossible, I'm not so sure.

Two days after I visit David Reich in his laboratory at Harvard, a party is held at the Cold Spring Harbor Laboratory, a world-class research institution on Long Island that was the site of the Eugenics Record Office until 1939. The celebration marks the ninetieth birthday of James Watson, one of the legends of twentieth-century genetics who, with Rosalind Franklin, Francis Crick, and Maurice Wilkins, helped discover the double-helix structure of DNA, for which Watson, Crick, and Wilkins won a Nobel Prize in 1962. Watson went on to become the laboratory's director in 1968 and was crucial in helping to get it funding over the years and building its reputation. A Grammy Award–winning pianist is invited to give a performance at the party, with no fewer than eight Nobel laureates among the four hundred guests.

Yet it has been known for years that Watson holds racist and sexist views. He was famously derogatory about his former colleague, Franklin, who did much of the experimental work that helped him make the discovery that led to his joint Nobel Prize. He also told the *Sunday Times* in 2007 that he was "inherently gloomy about the prospect of Africa," because "all our social policies are based on the fact that their intelligence is the same as ours—whereas all the testing says not really."

In 2010 David Reich witnessed James Watson's racism firsthand when they were both at the Cold Spring Harbor Laboratory for a workshop on genetics and human history. Watson sidled up to him and asked him something along the lines of, "When are you Jews going to figure out why are you guys are so smart?" Reich was appalled. Watson openly compared Jewish people to Brahmins, high-caste Indians who are known for being overrepresented in universities and high-status jobs. Traditionally they are India's educated, priestly class. Watson suggested that racial purity combined with millennia of selecting for scholarliness was the key to both Jewish and Brahmin success. He went on to make other racial slurs about Indians being servile, a trait he believed suited British colonizers, and about the Chinese, who he thought had been made genetically conformist by their society.

I wonder what Reich took away from this encounter. If, as Reich asserts, understanding the scientific facts makes it so impossibly difficult to be racist, how does James Watson manage it?

Reich hesitates. "Well, Watson is, you know, is probably more sexist than he is racist," he tells me awkwardly. "I don't know. I don't know. He's like uncontrollable. It's impossible to control Jim Watson. He purposely wants to create, to annoy people, to scandalize people, so I don't know. You can't control everybody. I do think that. So, yeah, I don't know."

There is a long pause, an uncomfortable half shrug.

"I just don't know."

CHAPTER 8

# Origin Stories

## *Why the scientific facts don't always matter*

The past is the problem.

Not just the politics of the past, the nineteenth-century theories of race and the colonial hierarchy of races that still permeate our subconscious, but the deeper past, further back. The problem is with how we build our ideas of who we are. When biologists try to understand ancient human migration, when they pick through our genomes and those of our distant ancestors, they are part of age-old efforts to piece together our origin stories.

In China, it's believed that taming the flooding of the Yellow River many thousands of years ago, by a man named Gun and his son Yu, marked the dawn of Chinese agricultural civilization. It's a legend that helps build national identity, serving a unifying purpose and lending a sense of superiority. Over the centuries, myths take on a life of their own, each generation recasting them to suit their needs until we can no longer tell the difference between myth and history. Before we know it, the glorious tales of our ancestors become our historical facts. Their ghosts become our icons. And of course we need to believe that they were better than they really were, that they were nothing less than superhuman. The founding myth of Rome is of the abandoned baby twins, Romulus and Remus, suckled by a she-wolf and rescued by a god. German nationalists told of a blond, rugged hero, Hermann, who defeated the Romans and united Germany's disparate tribes. These figures have become woven into national identity, pulling people together in the belief in the cosmic power of their forebears, cementing their particular claim on human civilization.

So science is not enough to forge identity. We also need stories to build a sense of who we are, even if the stories are held together with only threads of truth. There was indeed a German tribal chief who spearheaded a victory against the Romans two thousand years ago, on which the legend of Hermann is constructed. And in 2016 Chinese researchers confirmed that there really had been a giant flood around 1920 BCE. But legends must have been written around these facts until the people hearing them could no longer separate them from fact. The bloody realities became whitewashed over time, each iteration making the story cleaner, brighter, and more dramatic than the original. And this was necessary, not just for the sake of producing a gripping narrative but also because it's tough to build national pride and a sense of superiority around a dirty history.

The anthropologist Jennifer Raff, based at the University of Kansas, is, in her words, a "middle-class white girl" and she, too, grew up on a powerful origin story. "Those of us in the United States have been taught this idea of American exceptionalism," she says, "that our country is the greatest country, and is founded on these wonderful beliefs, this freedom and equality and democracy." It's a narrative that rests on the assumption that European pioneers in the seventeenth century filled a largely empty land with visions of a better society, deploying their unmatched skill and hard work to cultivate it. As in Australia, the indigenous inhabitants were framed as a dying race—if not gone, then definitely on their way out.

The subtext is that, without white Europeans, civilization couldn't have flourished in North America. The United States was *theirs* to make.

It's not easy to square this popular founding myth with the more brutal historical facts. Of course, Native Americans weren't as primitive, dwindling, or sparse as the settlers liked to portray them. When the land turned out to be less vacant than they hoped, European colonizers made every effort to empty it. Thousands of Native Americans died on the Trail of Tears, the forced relocation of several Native American tribes from their ancestral homelands in the southeastern United States to designated Indian territories west of the Mississippi, following legislation passed in 1830. Many more were killed by diseases brought by the migrants to which Native Americans had no resistance. Deaths frequently went unrecorded, which means today we have only the vaguest estimates of how populated North America really was before colonizers arrived. Genetic analysis published in 2011 suggests that the number of female Native Americans may have shrunk by half upon contact with Europeans five centuries ago.

We know this now. The founding myth becomes harder and harder to maintain. And yet it has strange ways of reasserting itself, even within academic circles, as Raff has found. Her work, trying to understand the distant past and the effects of race and migration, has shown her how easy it is for people, including respected scholars, to resist abandoning popular myths and racialized views of the past even in the face of undeniable evidence. Indeed the myth of American exceptionalism is so pervasive that an entire scientific theory exists to explain it, weaving in archaeology and anthropology with the notion that Europeans are the ultimate bearers of human progress. It's known as the Solutrean hypothesis.

Crafted in earnest in the 1970s, the Solutrean hypothesis takes its name from archaeological evidence of certain tool-making techniques belonging to the Solutrean culture, which existed in parts of what is now France and Spain between 23,000 and 18,000 years ago. The Solutrean method of making blades by forming long, narrow flakes appears to be similar to the technique used in New Mexico by a culture known as the Clovis, which is thought to be some 13,000 years old. If the Clovis tools, which would have been used to kill such beasts as mammoths and bison, weren't developed independently, then the Solutreans might have brought them to the Americas.

Geologists know that less than fifteen thousand years ago sea levels were low, allowing for a land bridge across the Bering Strait that would have joined modern-day Russia and Alaska. People could even have lived in the region between Asia and Alaska for an extended period of time before spreading farther eastward. According to new research, there may have been waves of migration in both directions, with some people returning to Asia. So far, however, the most convincing account of what happened is that the first Americans came from the west, not the east.

But if the Solutrean hypothesis is true, it means Europeans came to the Americas long before the colonists of the seventeenth century, that in fact they were among the first people to live here, having somehow crossed the treacherous Atlantic during the last ice age, which ended around 11,700 years ago. Way back then, vast swathes of the planet would have been covered in sheets of ice, and sailing—or perhaps snowboarding—3,700 miles across the Atlantic would have been a survival challenge of epic proportions. Yet those who defend this account believe that it was nevertheless possible, especially if there was a continuous ice shelf across the ocean, which could have provided fresh water and food throughout the journey.

It's a theory at the very margins of science, yet a small number of American archaeologists have staked their careers on it, publishing books on the

Solutrean hypothesis and clinging to the belief that more evidence will prove them right. Among the most vocal is Bruce Bradley, usually based at the University of Exeter in England. In the 1970s Bradley became aware of similarities between ancient stone tools dug up in northern Spain and those found in New Mexico. He couldn't believe that these similarities were coincidental.

In a telephone interview Bradley tells me, "The basic underlying technology, the way stone tools are made, unless you understand how many detailed choices you have when you're making stone tools, things seem like they could happen accidentally. . . . It's not just the blades, it's the way they made all the other tools. Virtually all of them have correspondences that are very, very striking between Solutrean and Clovis." This suggests that the peoples of the Americas must have reached there from Europe, possibly by traveling across the North Atlantic, via Greenland and Canada.

The political implications are clear. It could be read as a suggestion that Europeans had a prior claim to the Americas, because their ancestors were already here many thousands of years before Columbus arrived in 1492. When they came again, millennia later, then, they were only returning to a land that was already theirs. "I see it as intimately tied up with the idea of Manifest Destiny," explains Raff. This was a belief, particularly popular in the nineteenth century, that the European settlers who came to colonize what became the United States were somehow fated to do it, that it was written into the settlers' history before they even arrived. It's a narrative they thought gave them a moral claim to the land, and later helped to square the inhuman treatment and murder of Native Americans with the squeaky-clean founding values of the United States.

That said, evidence for the Solutrean hypothesis is thin, and getting thinner all the time. One of the glaring snags is that the two cultures, the Clovis and the Solutreans, existed so many thousands of years apart in time. Nobody has discovered any ancient bones in Europe belonging to people of Solutrean culture, only archaeological traces, such as the objects and art they left behind. So it's impossible as yet to connect modern-day Native Americans to Solutreans through their genomes.

Recent genetic evidence does show that almost all modern-day Native Americans have a shared lineage that can be reliably traced to people who once lived in eastern Siberia. The 12,700-year-old remains of a Clovis boy found in Montana have shown him to be more closely related to all indigenous American populations than to any other group. Raff explains that since ancient eastern Siberians were also related to the ancestors of modern-day East Asians, the obvious picture of migration is that the very earliest people to land in the Americas must have traveled through Asia and come by crossing the Pacific, not the Atlantic.

Then again, archaeology is a field in which it's difficult to ever be certain of anything. New evidence can emerge at any time, overturning everything people thought they knew about the past. Science more broadly almost always leaves room for doubt as well. Proving something definitively wrong is unusual because it requires you to look at every possibility in the universe, and rule each one out, one by one. Sometimes this can be done—the earth is round and it rotates around the sun, we know that for sure. But when it comes to studies of the past, it's notoriously difficult. There's always the chance that a skull will turn up from under a plowed field, or that a fresh scrap of archaeological evidence will bubble up from the Atlantic. This space for uncertainty, sometimes so small that you need a microscope to see it, is where the controversies live. And as far as Bruce Bradley is concerned, however controversial his Solutrean hypothesis, there's also a chance that he might be right, that history will dig up the evidence to vindicate him.

"Disproving is very, very difficult, and I don't even like the term[s] 'prove' and 'disprove,'" he says. "It's a matter of probabilities. Is this evidence more likely to indicate this than that? And that's the way we work all the time." Despite my efforts to put him at ease as we speak, he gets combative, occasionally even raising his voice during our interview. "I'm not trying to make anybody believe this hypothesis. I'm just putting the evidence out there and saying what we think it means."

Since he started working on this a couple of decades ago, Bradley has come under sustained criticism from fellow archaeologists and geneticists. One team of researchers in the United States even described his position as "Solutreanism," implying that Bradley and those who think like him had crossed the line from science into ideology. Jennifer Raff insists that the lack of evidence linking Solutrean culture to Native Americans is itself clear evidence. "You would expect to see a bunch of other technologies," she says. "You would expect to see cave art of the same kind, you would expect to see settlements, and you don't see any of that." The Harvard geneticist David Reich laughs when I ask him about the hypothesis. "The Solutrean hypothesis is a silly hypothesis. It's totally incompatible with the genetic data on so many levels," he tells me. "It's just not science. It's sharply contradicted by the science."

The geneticist Mark Thomas at University College London agrees that the theory has only the slightest likelihood, if any, of being correct. When I raise the subject with him, he sounds surprised that I'm even mentioning it at all. "Let's be clear," he says. "This is not a scientifically prevalent idea at all. If you are going to measure weight of argument in terms of word count, then maybe it seems prevalent. But this is very much like saying climate change is controversial because there are lots of words written saying there's no such thing as human-driven climate change. No." For Thomas, this is about a handful of researchers who have become so attached to an idea that they have embarked on a "confirmation bias odyssey," as he calls it, scouring the world for evidence while neglecting whatever doesn't fit.

Yet Solutrean hypothesis survives, popping up online, in scientific journals, and in the odd biology book—not just surviving, but also enjoying support from those with a vested interest in its being true.

| | | | | | | | | |

In the 2010 self-published novel *White Apocalypse*, white-skinned Solutreans, having crossed the Atlantic and settled in North America, are slaughtered by savages who later cross the Bering Strait and became today's nonwhite Native Americans. The author, Kyle Bristow, a Detroit lawyer active on the political far right in the United States, makes fictional heroes of real-life archaeologists like Bruce Bradley, painting them as victims of a conspiracy by Native Americans and liberals, who don't want to face the apparent truth that the original Americans were white.

Unsurprisingly, his book has become popular in white supremacist circles. One review stated, "This evidence could be the jolt whites need to awaken from our suicidal slumber." When it was republished in 2013, Bristow even included what were described as supplementary materials showing the validity of the Solutrean hypothesis.

For scholars such as Jennifer Raff, this comes as small surprise. The Solutrean hypothesis speaks to a nineteenth-century worldview that painted Europeans as the true inheritors of America, as the only ones capable of civilizing the continent. At the time, evidence of sophisticated technological cultures such as the Maya, the Inca, and the Aztecs, which existed long before the Europeans arrived, were only further fuel for confusion. "Since the beginning of the Americas, there has been a question: Who are the Native Americans?" Raff says. "People actually wondered, are they *humans*?

The first colonists did not really have a way to incorporate them into their biblical worldview. After their humanity was more or less accepted, it then became this idea that, well, are they responsible for creating the culture, the very sophisticated technologies and art and monumental architecture that we see?"

In the shock of uncovering complex ancient civilizations in the New World, the first Europeans imagined elaborate ways in which they could have gotten there. "It's so interesting to me when I look at ideas, alternative ideas, to explain the archaeology," Raff continues. The Solutrean hypothesis is just the latest iteration. "People are so desperate to find a non-mainstream answer to a lot of these issues. They won't just invoke Europeans, they will invoke aliens! They'll invoke people from Atlantis! Whatever they can find, as long as it's not Native Americans." The Book of Mormon, published in 1830, claimed that Native Americans were descendants of the lost tribes of Israel who migrated to the Americas around 600 BCE and had been cursed with a dark skin for slaughtering their righteous relatives.

Just because theories are exploited by the far right doesn't necessarily make them false. Raff is quick to add that although some supporters of the hypothesis may be motivated by racism, she doesn't believe that the researchers themselves are driven by this. "They are good scientists and they are legitimate scientists. They are very well respected." At the same time, though, she sees a doggedness about them that sets them apart. "Everybody I've talked to who actually knows them personally tells me that you cannot change their minds. Nothing will change their minds. Nothing. I wouldn't go so far as to call the Solutrean hypothesis pseudoscientific, exactly. It did start out as a legitimate area of investigation, but I see it right now as being almost more ideological. I mean, people are not accepting any evidence against it. If you're pro-Solutrean, that's it."

Of course, Bruce Bradley sees it differently. He tells me that Raff is "deluded, to put it bluntly." He believes he's been marginalized by the mainstream scientific consensus not because he is blindly clinging to a discredited idea but because he is brave enough to challenge the academic orthodoxy. As far as he's concerned, his detractors are the ones motivated by bias. "For me, it comes down to a lot of political stuff," he tells me. "When I first started promoting—not promoting, suggesting—this as a hypothesis, I was working in France and Spain, and different places over there. And I had very, very strong negative reactions from different people among colleagues in Spain. I think it's colonial guilt."

Bradley insists that his work is just good archaeology. "We've made it very clear all along that we're talking about our thing as a hypothesis," he tells me, sounding more than a little worn down. "People need to look up the definition of a hypothesis. A hypothesis cannot be right or wrong."

The debate intersects with another recent controversy, around one of the few ancient skeletons found in North America, known as Kennewick Man. Dated around 8,500 years old, when his middle-aged bones were discovered in 1996 by college students in Kennewick in the state of Washington, researchers were quick to spot that his skull didn't look particularly like that of other modern-day Native Americans. In fact, one archaeologist described it as looking "Caucasoid." Of course, one way someone with Caucasoid features could have ended up here all that time ago was if some form of the Solutrean hypothesis was indeed correct, that the ancestors of the Kennewick Man had traveled across the Atlantic from what is now Europe. A reconstruction of his face—cast in off-white, although nobody knew his real skin color—even weirdly resembled the English actor Patrick Stewart, best known for playing the captain of the *Starship Enterprise* in *Star Trek: The Next Generation*.

Meanwhile, local Native American tribes rushed to more plausibly claim him as their own, insisting that he must have been an ancestor or related to their ancestors. In their historical legends, the land on which they lived had been their home since the beginning of time. They were products of it, not migrants to it. According to this narrative Kennewick Man simply couldn't belong to anyone else, and so, having been dug up and manhandled, he deserved a proper burial, meaning one overseen by the tribes. This call to return the remains to the tribes seemed irrational and emotional to some in the research community. A bitter court battle began, pitting scientists against indigenous groups.

The struggle over the remains of Kennewick Man wasn't just about identity or ritual. It was also about unwrapping a dark and brutal history, in which the dreams of some people were used to override the dreams of others. In the nineteenth century Native American graves were often looted by anthropologists and hungry collectors, keen to claim their piece of this ancient culture before it disappeared, but with no respect for its traditions. These bones were rarely returned. The insults were not limited to artifacts and remains. As recently as 1990 blood samples from members of the Havasupai tribe, who have lived in the Grand Canyon for centuries, were taken by Arizona State University in the understanding that they would be used to study the people's risk of diabetes. In the end, without their permission, the samples were also used to study other medical and mental disorders, including schizophrenia. The university agreed to pay $700,000 in compensation.

So when Native Americans defended the bones of their ancestors, they were also standing up for the rights over their own bodies. Even so, in 2002 a judge finally ruled that the bones weren't necessarily related to any modern tribes, in large part because researchers had proposed that Kennewick Man didn't really look like the average Native American. As Kim TallBear in the Faculty of Native Studies at the University of Alberta has written, they privileged "genome knowledge claims over indigenous knowledge claims." Despite protests, scientists were given a green light to study the skeleton.

And then, slowly, came the revelations.

Researchers in Denmark, led by Eske Willerslev, a pioneer in population genetics and ancient DNA, revealed in 2015 that Kennewick Man was indeed more closely related to contemporary Native Americans than to any other group after all. The tightest genetic link was found to be to the local Confederated Tribes of the Colville Reservation, which had originally claimed him as an ancestor. The indigenous groups had been right all along. He was one of their own as much as it is possible to be when you are separated by millennia of time.

In February 2017, under legislation signed by President Barack Obama, Kennewick Man, now known by tribes as the Ancient One, was finally laid to rest in a traditional burial near the Columbia River. The act of rewriting the story with something closer to the truth, and then returning his remains to the tribes, carried layers of significance. "A wrong had finally been righted," a spokesperson for one of the confederated tribes told the *Seattle Times* when the Ancient One was buried. A fresh forensic reconstruction showed a face starkly different from the first. Like Britain's Cheddar Man, whose facial reconstruction went from white to black in the space of a century, Kennewick Man, too, was completely different the second time around. Now he was given long hair and dark skin. The resemblance to Patrick Stewart was gone.

It was a lesson in how much culture and politics can shape how people read scientific evidence. It's an easy mistake to project contemporary racial parameters onto the past, explains Deborah Bolnick, an anthropological geneticist based at the University of Connecticut. "If you see the genetic markers today that are found in western Europe, people will see those in the past and continue referring to them as western European, even if they're then also found in Siberia." It's another "indexing problem"—when the first available body of evidence influences subsequent thinking. Western researchers tend to have more access to European data because it's on their doorstep, so later discoveries elsewhere in the world are often interpreted relative to these. Bolnick tells me of the example of a skeleton of a four-year-old boy discovered in south-central Siberia and thought to have been buried there some 24,000 years ago. In 2013 this became the oldest modern human genome yet sequenced, and scientists learned that the boy shared some genetic variants with people in western Europe. "The way this got framed was: you have an individual in Siberia who has western European genetic markers, and so maybe this means that there was a migration from western Europe to Siberia," she says. In reality, the more parsimonious explanation, especially given the age of the skeleton, was that it was an east-to-west movement, not the other way. In other words, people in western Europe had *Siberian* genetic markers.

"Underlying assumptions and ideas definitely get embedded in ways that we don't even think about consciously, which can play out in the science," Bolnick adds. "Data don't say anything by themselves. We interpret data. We bring our perspectives, our framings to the data. You can use the same data to say many different things. I think modern genomic data provides the perfect example of that, because you can have different people who are all very smart and understand the data, who look at the same datasets and describe them in polar opposite directions."

Our stories get in the way of science.

It's impossible to escape our beliefs, our upbringing, our environment, even the pressure of wanting to be correct when it comes to interpreting the facts. Romila Thapar, an Indian historian, writes, "In contemporary times we not only reconstruct the past but we also use it to give legitimacy to the way in which we order our own society." Jennifer Raff believes this is quite clearly in play when it comes to the Solutrean hypothesis, just as it may have been when it came to hypotheses regarding the ancestry of Kennewick Man when he was discovered. There are powerful reasons why researchers want to believe their own story is right, even when evidence declares otherwise.

The past is always at the mercy of the present.

| | | | | | | | |

"I remember during the Yugoslav Wars, I was in Paris."

Kristian Kristiansen, a senior professor of archaeology at the University of Gothenburg in Sweden, works with ancient-DNA expert, Eske Willerslev, who carried out the investigation into Kennewick Man. Kristiansen is infectiously enthusiastic about this powerful new field of science, but as a longtime archaeologist, he also has a measured perspective on the past. He agrees that leaps in genetics have the power to overturn everything we thought we know about ourselves. They certainly challenge racial stereotypes by showing us just how much we have always mixed together throughout the past, and how much we have in common. But at the same time, he warns from his own personal experience that the political power of such insights have their limits.

In the nineties he saw this play out for himself, during violent ethnic conflicts between Serbs and Croats in the struggle for independence in the former Yugoslavia. "I was a visiting researcher and I was living in Paris together with some expelled, you could say, archaeologists from Yugoslavia," he says. The lives of ordinary people meant little in that place at that time. In the push for territory, a program of ethnic cleansing, mainly of Bosnian Muslims, led to hundreds of thousands of civilians being forcibly displaced, women being systematically raped, and murders so numerous and methodical taking place that they rose to the category of genocide. Political leaders deliberately rewrote history to cast some ethnic groups as having a claim to certain tracts of land.

Reputable historians and archaeologists found themselves fighting an intellectual war against nationalist ideologues who wanted to justify their actions by promoting false versions of the past that suited their cause. "And nobody wanted to listen," says Kristiansen. "That was the shocking thing. Nobody listened. And they published in newspapers; they did everything they could to get it across, but in the heat of the whole thing, they failed." When push came to shove, truth became victim to politics. The facts mattered only if they suited the power-hungry agenda. "Suddenly things can turn from left to right in a split second when politics changes," he says, clicking his tongue.

This wasn't anything new. It had been seen before, most obviously earlier that same century when the Nazis pulled together whatever racial theories they could to defend the genocide of millions of Jewish people and members of other groups during the Holocaust. Then, too, mainstream scientists and archaeologists found themselves marginalized and sacked while those whose ideas favored the regime found themselves promoted and celebrated.

Bettina Arnold, a historian and professor of anthropology at the University of Wisconsin, Milwaukee, has researched just how gross these intellectual abuses were in the years leading up to World War II. After their country's humiliating defeat in the First World War, many Germans were looking for ways to rehabilitate their national pride, and the search for a more glorious prehistory was one means to that end. By promising to mend this collective feeling of bruised self-respect, the Nazis managed to gather public support. Slowly, they harnessed archaeological evidence that fit their account of a great "Germanic" past. At the same time, by proving that the ancient roots of the German people were to be found across Europe, they could lay moral claim to territory beyond their own borders. In their minds, they would expand to form an empire based on the original Germanic race, which they believed itself originally stemmed from noble, light-skinned Aryans and was physically and mentally superior to all others.

Their intellectual framework came partly from the linguist Gustaf Kossinna, who had been appointed a professor at the University of Berlin in 1902 and went on to become one of the country's most influential thinkers. By the time the Nazis came to power, Kossinna was dead. But the Third Reich had already nurtured his theories, seizing upon his argument that culture and ethnicity were wrapped up in each other. His ideas implied that when archaeologists uncovered evidence of shifting cultures, they were also seeing evidence of migration. So if they could find archaeological proof that the same cultures they could see in Germany had existed elsewhere, this would also be proof that ancient Germans had lived there, that this was also part of *their* rightful territory. Archaeology, folklore, and anthropology combined in service of this political idea.

The Nazis, says Arnold, were bent on "proving that there was some kind of, well we would say, genetic racial essentially commonality." It was about expanding the boundaries of the traditional homeland using race as a rationale. This is not to say that this idea was welcomed, or even widely accepted. Kossinna was heavily criticized within his own lifetime for the quality of his work, most notably "for the kind of cherry picking that he engaged in," she adds. "You pick certain parts of material culture that support your arguments, you ignore those that don't. This is obviously a danger anyway in any archaeological interpretation. In his case, it was quite easy to pick holes in the arguments that he was making, and people did. Even his contemporaries did."

Part of the reason that Kossinna was drawn to the Nazi Party as they were beginning to claw their way to power in the early twentieth century is that in them and their ideology he found support he didn't necessarily have from his peers, mainstream historians and archaeologists. "He was a bit of a marginal figure initially early on in his career," Arnold says. "He had been rejected by sort of the mainstream cultural historians of the day. He had a hard time finding an academic job. There is a lot of the personal bitterness." But in the Nazi Party "he found a niche and a place where he could matter, where his work was accepted and seen as important." Kossinna wasn't working for the party when he first developed his theories, but he was certainly motivated by an ethnically charged worldview that became useful later on. By the end, the party turned him into an icon, a founding father for the regime.

The politics suited him as much as he suited the politics. All the way up to his death in 1931, Gustaf Kossinna was firmly on board with Nazi ideology. Many of his publications make clear that he was aware of the political ramifications of the research he was conducting, says Arnold: "He fully supported the idea that archaeology should be a handmaiden of the state." Arnold has noted that in the first two years after Adolf Hitler came to power, eight new academic chairs were created in German prehistory. History was deliberately rewritten and appropriated by the party. The infamous swastika we associate with the Third Reich was employed after German archaeologists found the same prehistoric symbol on old German pottery. The "SS" double lightning bolt that featured on Nazi uniforms was similarly adapted from an old Germanic rune.

Everything was recast through a political prism. Archaeologists writing for mainstream journals were replaced by those who toed the party line, and Germanic cultural influence on Western civilization was intentionally exaggerated. In one bizarre instance of wishful thinking, the ancient Greeks were painted as ethnic Germans who had long ago somehow survived a natural catastrophe before developing a sophisticated culture of their own in southern Europe. The German tribal chief Hermann, who had defeated the Romans in the Battle of the Teutoburg Forest almost two thousand years earlier, was dragged into service, too. Under the Nazis, his statue—erected in the forest in the nineteenth century—became a focal point for nationalist pride, a reminder of a golden age of heroism.

Gustaf Kossinna remains a cautionary figure for archaeologists, as he does for academics more widely. The problem throughout, Arnold argues, is that archaeology—with its shortage of evidence and abundance of interpretation—has always lent itself to misinterpretation. The same may be said of other scientific fields, especially when data is thin on the ground and there are plenty of people desperate to speculate on the meaning and significance of what little there is. This has certainly been the problem with race science, and the study of human variation.

Kossinna is a reminder that shaping evidence around ideology, selecting specific results to suit a narrative, or even just failure to exercise care when it comes to interpreting or presenting data can lead to disaster. Arguably, what Kossinna did was no different from how scientific information was manipulated by anti-abolitionists in the American South in the nineteenth century or by British imperialists who made the case for colonial rule by framing themselves as racially superior. But today it's a lesson taken seriously in Germany. When the two-thousandth anniversary of the Battle of the Teutoburg Forest rolled around in 2009, the celebrations were sober, with a marked shortage of volunteers wanting to play the role of the Germans in a battlefield reenactment. Most volunteers wanted to be the Romans. Showing a distinct lack of nationalist fervor, even a spokesperson for the local museum told a reporter, "I hope people in the future will take a closer look at history, question what they have learned, and review the sources."

"There is in Germany among my fellow archaeologists a really high sensitivity towards political misuse, towards simplification," says Kristiansen, "because they have seen the way that the Nazi regime constructed a false prehistory, by taking elements of the established prehistory and then twisting them." For example, in 2015 the Harvard geneticist David Reich was working on a paper for publication in which he examined evidence of the very same prehistoric culture that Kossinna had once described as Germanic. A German archaeologist who had supplied the team with skeletal samples was so concerned about the same conclusions being reached about links between migration and cultural change that the Nazis had made that he and a number of other colleagues asked that their names be taken off the list of authors.

There is good reason to be cautious. In spring 2018 the prestigious science journal *Nature* issued an unusual editorial stating that, reminiscent of Gustaf Kossinna, "Scholars are anxious because extremists are scrutinizing the results of ancient-DNA studies and trying to use them for similar misleading ends." It was the kind of warning that would have been unthinkable in a scientific journal a decade ago. "They worry that DNA studies of groups described as Franks or Anglo-Saxons or Vikings will reify them." The suggestion was that people out there are actively abusing science for racist purposes. In 2018 the *New York Times* reported that white nationalists had been seen "chugging milk" at gatherings to demonstrate a genetic adaptation shared by many Europeans that allows adults to digest milk (a trait, incidentally, common to many nonwhite populations, too, who have historically also kept dairy cattle).

Kristiansen has witnessed this kind of racially motivated cherry picking and distortion of scientific data. "Every time we publish, it goes into the global database. And what we can see is a lot of people are sitting out there that have all kinds of blogs where they go in and reanalyze data and see if they can falsify or get other results." He suspects some of them may be fellow academics, but others seem to be enthusiastic amateurs. From what he can tell, they are deliberately scouring the genetic and archaeological data for evidence that fits in with their pet political or racial theories. In one memorable incident he was drawn into email correspondence with a respectable Canadian sociologist with a professorship at a public university who had cited his research. As they emailed each other, it slowly became clear to him that this man had views sympathetic to white supremacists.

"Everything can be twisted," Kristiansen warns me. "Everything."

| | | | | | | | |

In the spring of 2018 a smattering of news reports began circulating in the Indian media, which could have been lifted straight from Germany in the 1930s: the Indian government had set up a committee to rewrite history.

According to the reports, this was a decision that threatened to slam headlong against established scientific and historical facts, promoting a mythical version of history that painted India's dominant faith, Hinduism, as being central to its entire past. This particular origin story had been around for a century or so, enjoying varying levels of support among the populace, but it had become increasingly popular in recent decades, especially with the election of a conservative Hindu nationalist government in 2014. Now, it seemed, identity politics was being ratcheted into a higher gear.

Appointed by the prime minister, the twelve people on this new committee included a former senior official with the Archaeological Survey of India and the minister of culture, who was apparently keen to introduce a "Hindu first" account of history into schools. Established facts about evolution, migration, and genetics would be thrown out of the window in favor of a firmly religious narrative, one insisting that ancient Hindu texts are fact, not myths, and that those of other faiths have no claim to India.

In her book on contemporary identities in India, *The Past as Present*, the historian Romila Thapar explains that the idea of a Hindu homeland has its roots in the struggle against British colonialism and efforts to construct a new national identity once independence was won. Just like in Germany following the First World War, politically motivated accounts of Hindu superiority have offered ordinary Indians an opportunity to reclaim their self-respect and assert some collective pride. But in the process, India's ancient past, which is far from fully documented, has become a tool for projecting notions of technological and cultural superiority. Some of the members of the government committee to rewrite history, like other religious nationalists, believe that India belongs only to Hindus, even going so far as to suggest that Indians have no ancestry anywhere else, not even in Africa, where our species originated. One member, a Sanskrit scholar, reportedly believes that Hindu culture is millions of years old, an order of magnitude older than our human species.

For Hindu nationalists, their ancestry and religion both tie them deeply to their land. Some have absorbed old European and American theories of an ancient, noble, pure-blooded Aryan race, and claim that these Aryans did indeed originate in India, living in the sophisticated cities of the Indus Valley Civilization in northwestern India thousands of years ago. Like the way the Nazis saw themselves, they see modern-day Hindus, particularly light-skinned, higher-caste Hindus living mainly in northern India, as direct descendants of the Aryans. It is a connection that is thought to be timeless, but also that makes them superior to everyone else on the planet. As Thapar writes, these ideologues believe that "the Aryans of India were not only indigenous but were the fountainhead of world civilization, and that all the achievements of human society had their origins in India and travelled out from India."

It's a version of history that doesn't withstand much intellectual pressure. The oldest settlements to have been excavated in India, belonging to the Indus Valley Civilization of 3,000 to 1,000 BCE, confirm that the modern-day Indian population must be a result of different waves of migration, some more genetically related to Europeans, others less so, and everyone a mix. Hinduism and its cultures, too, have changed through time, and according to Thapar, bear little relation to the earliest civilizations.

But these facts don't always seem to matter, I'm told by Subir Sinha, a researcher at the School of Oriental and African Studies in London, who has been tracking the rise of extreme Hindu nationalism over the years. "All I can say is that people who used to be scientific and rational at one time will now take a view that this [account of Hindu origins] is possible."

While certain facts are deliberately ignored, at the same time there is a desperate desire to find others that do fit the ideology. The parallels with Gustaf Kossinna and Nazi Germany are striking. For example, Indian archaeologists have been tasked with digging up evidence of places, people, and events described in Hinduism's ancient texts. The legends include tales of demons, flying machines, and monkey-headed and elephant gods. Some nationalists say these weren't just beautiful allegories, but hard historical details. One example Sinha gives me is that of the mystical river known as the Saraswati, which is at the center of much of the action in one of Hinduism's scriptures. "One of the first things the government did when it came to power this time was to set up a task force to identify the Saraswati River," he tells me. "There is a kind of will to truth. We will make this to be the truth if we try hard enough."

Not just history and archaeology, but biology, too, must be deployed to support the myth. "They care a lot about what the scientists are doing," adds Sinha. When geneticists release new findings about human ancestry that don't sit well with the religious narrative, they are seen to pose an intellectual threat. It's a problem that has already landed at the door of Kumarasamy Thangaraj, India's leading population geneticist, based at the Centre for Cellular and Molecular Biology in Hyderabad. As one of the scientists who helped prove that modern Indians are the product of repeated migrations, he is well aware of the controversies surrounding the work of researchers like him. When he carried out work on Indian population genetics in collaboration with international colleagues, they deliberately decided to describe ancient migrant populations not as being African, Iranian, or Middle

Eastern in origin, as they might have for accuracy's sake. Instead, they called them "ancestral north Indian" and "ancestral south Indian," in an effort to be politically sensitive. With this wording, they avoided upsetting those who believe that Hindus spontaneously originated in India.

"It has not come to that level where I have to argue with them. People talk to themselves. They never fight back to me or oppose my findings, but that exists," Thangaraj tells me, diplomatically.

Even so, Romila Thapar, Subir Sinha, and other academics have expressed strong concerns about what they see happening in India. "Most of the politics of connections with land and nature and 'we are the true people' tends to be of a fascist, right-wing variety," Sinha explains. "They believe that civilizations should be based on a true homeland of righteous people, which have the same religion and language." Religious minorities, particularly Muslims, have been picked out for persecution in this increasingly charged political environment. The worst incident so far took place in early 2018, when Asifa Bano, an eight-year-old Muslim girl living in the Indian-administered part of the state of Kashmir in north India, was taken to a Hindu temple and gang-raped over a period of days before she was murdered and her dead body dumped in a forest. Two government ministers attended a rally in support of the men accused of the crime.

The barbarity of how nationalism plays out may make it feel as though facts are peripheral, that they don't really matter, not when lives are at stake, not when young girls can be gang-raped for their faith. But for the religious nationalists, says Sinha, "the past matters a lot, for them to be confident in making the claims of greatness they want to make, claims to greatness in the world but also claims to land, power, claims to the right to show down people of other religions." It gives them privilege over the truth, a version of their social structure that they can then sell to others. It throws weight behind the fists, it gives people the sense that what they're doing is morally justified, because this is the order of things as they see it. This is how the world was created, and they are only bringing some of that order back to chaos.

The nationalists must turn to the past for reassurance. The past is their problem. But then, arguably, so it is for all of us, in smaller and bigger ways. When we study our genetic ancestry, aren't we also looking for clues about who we are, trying to reaffirm a story we have of ourselves? Why does it matter to some people that their ancestors were Vikings or Egyptian pharaohs? Does being related to Napoleon or Genghis Khan make a person living today any different from the next person? When we claim ethnic or racial pride, what are we doing but trying to piggyback off the achievements of those who went before us? It's not enough to be who we are now, to be good human beings in the present. The power of nationalism is that it calls to the part of us that doesn't want to be ordinary. People like to believe that they are descended from greatness, that they have been genetically endowed with something special, something passed down to them over the generations.

CHAPTER 9

# Caste

## *Are some races smarter than others?*

On a smoggy January day I set out from the Indian capital, New Delhi, for the nearby state of Punjab to visit my extended family. It's a journey I've taken many times before, usually napping my way past the lush farmland and fruit sellers flanking the highway, stopping only at a *dhaba* for a butter-soaked lunch before returning to sleep. But this time I stay awake to study the faces I pass while I'm on the road, to watch the bodies jostling through the traffic.

I stare and I compare. The word "brown" doesn't do any of us justice. Every possible skin color is represented: ebony, paper white, yellowish, and countless other shades, along with almost every possible feature. India is unique. The sheer size of the country and its environmental variety, from the sun-drenched beaches of the south to the snowy Himalayas of the north, seem to be paralleled by the physical diversity of its people.

An encounter the same morning had made me look with fresh eyes at this place I thought I already knew. I'd met Sridhar Sivasubbu, a geneticist, in his scrupulously tidy office at the Institute of Genomics and Integrative Biology on Mathura Road in the heart of the city. Part of Sivasubbu's work is to investigate human genetic variation within India, with the aim of battling rare diseases. In a nation of more than a billion, he told me, no rare genetic illness is actually all that unusual. With this many people, rarity becomes a relative concept. But what fascinates him more than that is the variety. India is a microcosm, an entire hemisphere represented inside one country.

"We have something like fifty-five populations, major populations. Then there are minor populations within the country, and five linguistic groups are there," he explained. Regions overlap genetically with other nations in

South Asia and parts of the Middle East. The Andaman Islanders have genetic affinities to Aboriginal Australians. This breadth of difference may be one reason why today India is the only country left in the world to have its own government-funded Anthropological Survey, designed to study the biological and cultural variation of its citizens.

But there's also more to all this wondrous human diversity than meets the eye. One of the unsettling reasons that Indians exhibit such physical difference is that it's partly self-imposed—the culture demands it. Many centuries of marriage within fiercely tight-knit communities and a caste system that stretches back perhaps two millennia to keep privileged and nonprivileged people apart, reinforced by the British under empire, have deliberately maintained the separation of populations.

Romila Thapar has noted that there was always intermixing between groups in India, that the strict divisions today probably weren't as strict in the past. But unlike neighboring China, which though larger than India is not quite so ethnically diverse, in India, freedom to marry and move between groups does seem to have been restrained. Millions prefer to marry within their own religion, color, caste, and community, however shallow a pool of potential partners that might leave them with. And their preferences are policed not by authorities enforcing laws but by families: cases of couples being attacked or killed for falling in love inappropriately occur on a regular basis. Intercaste marriage was legalized in 1954, yet a survey in 2016 found that as many as 40 percent of adults in Delhi who didn't belong to the lowest castes thought there should still be laws preventing intercaste marriage.

"You can find this in Hindus, you can find this in Christians, you can find this across India," said Sivasubbu. "So it's not about religion. It is about customs and marriage practices that have been passed down over generations. Indians tend to marry within the larger community that they live in, and in spite of all the few hundred years of knowledge that we have acquired, we still follow conservative and traditional marriage practices."

My mother grew up with these values. Although she ended up married to a man of a different caste, religion, and community, she nevertheless has a romantically fatalistic view of life, steered by a society that has forever told her that everything is circumscribed, that has for hundreds of years kept people in their place. For those raised this way, society's hierarchies feel knitted into their bodies. Their faith in the power of heredity is so strong that it overwhelms how they think even about themselves. They find nothing odd in a dynasty that runs over multiple generations, be it political, artistic, or commercial. Some extol the virtues of caste as giving everyone a valued place in society, forgetting that for people consigned to a life of cleaning toilets, it's little solace to be told that this is where they belong in the cosmic scheme of things.

Parallels have been drawn with the British class system, or race in the United States, but caste has features of both and of neither. It is ugly in its own way. At birth, you inherit your place in this social hierarchy, and few transcend it. An "untouchable" at the bottom will be given society's dirtiest work, existing in a permanent state of impurity, while those at the top of the ladder are a kind of aristocracy, favored for jobs and education (people I meet as a journalist still like to drop into conversation that they are Brahmins, the highest, priestly caste, expecting this to carry some weight). The distant origins of all this are likely to have been strategic in part, to keep wealth and property within families. Some castes tend to be congruent with certain trades, creating generations of teachers, merchants, or fisherman. Those at the peak are apparently supposed to show benevolence to those below, but in reality there is considerable discrimination and violence.

Successive governments have brought in reservation policies, setting aside jobs and scholarships for disadvantaged castes. Even so, a report in 2014 by the international advocacy group Human Rights Watch found that teachers in some schools were still forcing lower-caste students to clean toilets and sit apart from everyone else. Top universities and colleges admit almost exclusively students of higher castes.

When the Nobel Prize–winning geneticist James Watson compared Brahmins to Jews by claiming that both had been bred for academic excellence, he was using the language of both caste and science, seeing the two groups' behavioral characteristics as genetic qualities passed down over centuries. Their fortune is believed to lie beyond circumstance, to be part of what they are. As a famous Indian economist belonging to a higher caste once told me, this is Indian culture, and it is unrealistic to expect change. It is taken as given that people are fundamentally different, that they are born a certain way. Everyone becomes trapped in the net of their ancestral history.

So Sivasubbu's comments overshadowed my thoughts as I traveled to Punjab. In India, skin color doesn't always faithfully betray someone's caste, but there's a lingering view that being fairer is better, that it denotes a higher status. The four main caste groups are sometimes even designated by color—white, red, yellow and black—not unlike the way Europeans classified human races in the nineteenth century. "Somebody very tall or very fair-colored, obviously they would select an individual with similar features," I had been told matter-of-factly by Indian population geneticist Kumarasamy Thangaraj. "So there is a selection operating, not by natural but by man-made selection." I recall when I was flat hunting in Delhi before starting my first job there as a reporter, I was asked to list my skin shade on the rental form. Having only thought of myself as "brown" for most of my life, I had no idea what to write. The letting agent took a good look at me, and with a dirty smile scribbled, "Wheatish." Color takes on a new subtlety when every degree of pigmentation matters.

In both biological and social terms, then, India has long been a unique case study for scientists. This is why the race scientist and eugenicist Reginald Ruggles Gates was so captivated by the country. Here, systematic discrimination, the notion that groups of people are biologically pure and should be kept separate, that some populations are different from others, isn't just an ideology. It's a living practice.

| | | | | | | | |

In the 1950s the Indian geneticist L. D. Sanghvi wrote that people in his country "are almost under an experimental environment . . . broken up into a large number of mutually exclusive groups, whose members are forbidden, by an inexorable social law, to marry outside their own group."

When the idea of race research became unpalatable after World War II, Sanghvi was one of the first to turn to population genetics. In India scientists like him could explore what happens when human groups stay "pure"—the kind of purity that nineteenth-century race scientists imagined might be possible, that Hitler wanted to see in Germany for the Nordic "master race," and that white supremacists still want to see in parts of Europe and the United States. The grand social experiment that had already taken place in Indian society could reveal firsthand how the world would look if people mated only inside their narrow communities, whereby certain qualities and skills were selected for over many generations.

What is remarkable is just how widely Indians today believe that caste is deeply, biologically meaningful, that it has created exactly what it must have been intended to create: a social order reflected in biology, with the smartest and most gifted at the top, and others in various professions with their own skills below, as warriors and merchants, or cleaners and servants. Even scientists think this way. "The caste system, whatever has been practiced for last several generations, or several thousands of years, has a definite impact on everything," Thangaraj told me. Character traits and abilities get passed down over generations, he implied. "The offspring that's coming out of that founder are going to have such character. Usually, then, they become very unique features of that particular population."

Sridhar Sivasubbu, too, suggested to me that people are biologically suited to the groups they are born into. By separating themselves for so long they had created genetic enclaves with particular talents, making caste not only a social reality but a biological one, too. "Clearly certain communities have certain biological abilities which they are born with. . . . We all look different and we each have our own strengths and abilities," he said. For him, the differences are so profound that castes are analogous to separate races, as the population geneticist Luigi Luca Cavalli-Sforza might have defined them. "You could call two groups, two peoples, as completely different races and treat them as two separate entities. Or you could just celebrate both of them and say that they are different and each has the unique strength and weaknesses."

By way of an example, Sivasubbu pointed to Haryana, one of the states bordering New Delhi, which happens to be home to a disproportionate number of athletes, particularly wrestlers and runners. "So, clearly they seem to have a better physique in terms of strength," he suggested. Another example he offered was that of several tribal communities, which he claimed were naturally gifted at archery.

This casual speculation surprised me, coming as it did from a respectable geneticist. It showed that more than half a century of research into human variation hasn't eliminated prejudice within science, wherever it's done. Old stereotypes are still being projected onto people, but perhaps in new ways. In Haryana there is certainly a long cultural tradition of wrestling in some families, for which people train their whole lives. But lifelong training could just as easily explain the prevalence of athletes as any innate ability. And if tribal communities happen to contain more skilled archers, this is most probably because they have traditionally been the ones to use bows and arrows, developing their skills through sheer hours of practice.

There's a slippery slope here, assuming that everyone in a particular community should be limited to certain paths in life. Indeed, social categories like these were harnessed and promoted by the British during colonial rule. My father's family, who were in the military and fought in both World

Wars for the British Empire, were designated as one of the "martial races." They were seen physically and morally as the perfect soldiers. My father became an engineer, his brother a headmaster. Few of his siblings followed the family tradition, and certainly none of the children. We are not all natural warriors.

Yet I can't help but ask myself whether Sivasubbu has a point. In 2018 scientists were amazed to discover that the nomadic Bajau people of Southeast Asia, who live almost entirely at sea, surviving by free-diving to hunt fish, had evolved an extraordinary ability to hold their breath underwater for long periods of time. The Bajau tend to have disproportionately larger spleens than nearby populations of farmers, which possibly helps them to keep up their blood oxygen levels when diving. There appears to be a measurable genetic difference between them and others, sharpened over many generations by living in an unusual environment.

This raises a question we don't like to ask out loud, but one that goes to the heart of the race debate. It is where race science began, with a belief that neglects history and jumps straight to the conclusion that the human zoo is like an animal zoo, each of us defined deep down by our stripes and spots. And it leads straight from the offensive observation made by James Watson on the preponderance of Jewish intellectuals and Indian Brahmins in academia. Might it be possible, as Watson implied, for a group of people, isolated enough by time, space, or culture, to adapt to their particular environment or circumstances in different ways? That they could evolve certain characteristics or abilities, that they might differ in their innate capacities?

Wandering down this road may be of some scientific value, but it is also risky, I know, and it's paved in blood. "Could there be psychological differences between population groups?" I asked Thangaraj, tentatively. I go further. "Differences in cognitive abilities?"

"That kind of thing is not known yet," he replied. "But I'm sure that everything has genetic basis."

| | | | | | | | |

Back in London, I'm on the train on my way to a leafy corner in the south of the city, Denmark Hill.

The question of whether cognitive abilities, in the same way as skin color and height, have a genetic basis is one of the most controversial in human biology. It's a grenade. And Robert Plomin is one of the few who has dared to handle it. A professor of behavioral genetics at the southern campus of King's College London, he has dedicated his career to the search for the roots of intelligence, becoming one of the most divisive researchers in mainstream science. His work has far-reaching implications for how we think about human difference.

Tall, with a smart white beard and a crisp, pale blue shirt, Plomin is disarmingly charming in person, and rarely says anything to me that doesn't sound perfectly reasonable. I find him to be ruthlessly careful with his words. But there's a message in his subtext. He moved to Britain from the United States in 1994, becoming known for brandishing the view that the cognitive differences between individuals can be accounted for largely in some way or another by genetics. The implication is that we are who we are, however we're raised. But at the sharpest end, a few go so far as to take it to mean that the achievement gaps we see between population groups (or races, as some might call them) may also just lie in their DNA.

It was in the 1970s, working at the University of Texas at Austin, that Plomin decided to dip his toe into the controversial field of behavioral genetics, eventually asking himself whether individual differences in intelligence might be heritable and to what degree. It was perilous scientific territory with enormous social implications.

For him as a psychologist, he tells me, "It was still kind of forbidden to study genetics. It was dangerous professionally." Part of the reason, of course, was the dark history of eugenics in the United States and Germany, which saw people sterilized or killed in the belief that they would pass on their "feeblemindedness" to their children. The handful who doggedly stuck with intelligence research after the war also tended to say inflammatory things. For example, in 1969 the American educational psychologist Arthur Jensen claimed that black Americans had substantially lower IQs than white Americans, and that IQ was also significantly heritable. This in turn suggested that the apparent black-white intelligence gap wasn't because blacks were socioeconomically worse off or discriminated against. Plomin tells me he both personally knew and defended Jensen against his critics before he died.

At the time, there was no genetic evidence to support Jensen. But Jensen predicted that scientists would one day discover such things as "intelligence genes," which he believed would be "found in populations in different proportions, somewhat like the distribution of blood types." To compare "intelligence genes" to blood types was telling because it was already known that blood types do vary in frequency between population groups. This would lead neatly to his next bit of speculation: that fewer of these "intelligence genes" would be present in the black population than in the white.

Jensen's work went on to feature prominently in *The Bell Curve*, the controversial 1994 book by Herrnstein and Murray. Herrnstein, a psychologist like Jensen, had long insisted that intelligence was heavily heritable, warning in 1971 that America was already slipping into a genetic caste system based on intelligence in which successful people were marrying each other and creating more successful children, while the unemployed languished because of their intellectual disadvantages, which "may run in the genes of a family just as certainly as bad teeth do now." Parallels with India's historic caste system were clear, except here it was being framed in purely biological terms. In his vision of the world, class, wealth, and race overlapped because of biology, not because of history.

In the United States, much heat still surrounds the nation's apparent black-white IQ gap. In 2006 the Canadian psychologist John Phillippe Rushton, who had been a pivotal figure in the Pioneer Fund and the *Mankind Quarterly*, published a brief commentary together with Arthur Jensen in the journal *Psychological Science* in which they repeated what they and Herrnstein had insisted their whole lives: that there was a racial gap of around fifteen IQ points and that the evidence it was innate was unassailable.

As they knew, proving their hypothesis meant separating the effects of nature and nurture. This is the challenge that has dogged human biology for more than a century. How do we know that differences between people are genetic or innate in some other way, and not just the product of social and cultural factors? Every human being is a unique product of biology and environment, which makes answering this question almost impossible. Without being able to raise human genetic clones in a laboratory to run experiments on, the backbone of the psychological efforts therefore became twin studies. By observing the similarities between identical twins, researchers have for decades thought they might be able to discern whether certain traits might be more heritable than others. But twin studies, too, are tainted by a toxic past. Josef Mengele, the notorious Nazi doctor who trawled concentration camps for involuntary subjects, had picked out young twins to deliberately mutilate (via amputations) and dissect. He carried out on humans the kind of research that, for ethical reasons, is done only on flies. And scientists learned nothing from it except just how far into hell a person would go to get his results.

For decades after the war, most researchers were wise to leave twins alone. In 1979 a psychologist at the University of Minnesota, Thomas Bouchard, reignited the flame. Studying a hundred pairs of twins who had been separated in infancy and then raised apart, he estimated that genetic factors accounted for approximately 70 percent of their variance in IQ, implying that the bulk of the differences we see between people who are healthy and well cared for are decided at birth. The rest lies in other factors in the environment, such as upbringing and schooling. Bouchard and his colleagues speculated that, at least in middle-class families living in industrialized societies, "although parents may be able to affect their children's rate of cognitive skill acquisition, they may have relatively little influence on the ultimate level attained." The implications of Bouchard's work, which had been financed initially by the Pioneer Fund, were obvious. If some people weren't doing so well at school—black American children, for example—it was nobody's fault but their own genes'.

At the time Bouchard was vilified, picketed, and called a racist. Many decades have passed since then, but in some ways Robert Plomin has inherited this tarnished mantle. He has become part of the push to rehabilitate intelligence research, running his own studies using twin and sibling data to understand inheritance. The key is to see, as Bouchard did, whether twins raised in separate environments end up the same. And he believes that, more or less, they do.

| | | | | | | | | |

"Everything is heritable," Plomin tells me straight out.

"In fact, I am not aware of anything reliably measured that's been shown not to be heritable in terms of psychology. . . . Everything is moderately heritable." Using studies of twins—particularly those who have been adopted into different families—his estimate for the portion of intelligence that is heritable is around 50 percent, which may be far lower than the figure Bouchard came up with, but is still pretty high. If half of our intelligence can be decided by our genes, then a large part of academic achievement may well be innate, immutable.

There are some important caveats, however. First, measurement of intelligence is itself fraught with pitfalls—nobody is fully satisfied that any IQ test can really do the job. Second, rates of heritability aren't the same for everyone. They depend critically on the environments of the people you're studying. Take a packet of seeds and shake half of them into a container filled with nutrient-rich soil, blessed with all the water and sunshine they need. Take the other half and put them in a container of poor soil with little water and light. In both pots, individual plants will grow to various heights, some taller, some smaller. The differences you see *within* each pot are highly hereditary because their conditions are the same. But in the first pot, each seed has been given the full opportunity to achieve its potential. In the second container, they haven't been, so the plants in this second pot will inevitably look smaller and scrappier. In this weaker one, even the naturally strongest seed may not reach the same height as many of the plants in the more fortunate container. So the differences *between* the pots are not attributable to heredity.

Some traits, such as hair color, are very strongly determined genetically. Hair color doesn't change depending on the environment, unless perhaps that environment is a hair salon. Even skin color is to some extent affected by how we live. A group of paler-skinned children who play out in the sun all day will temporarily end up with darker pigmentation than paler children who don't, but this difference is purely due to environmental factors. So when scientists say that a trait such as height or intelligence is partly heritable, the only way they can know how much is by looking at people in the same normal, healthy environments, with few differences in how they are raised or treated by society. If some of the people are deprived, it can obscure the genetic influence.

For instance, studies have shown that although the two Korean populations were until recently the same, today, North Koreans are on average a little shorter than South Koreans. An alien landing on our planet with no knowledge of their histories might call it a racial difference, but it is purely down to their dramatically different economic circumstances. Height has very high heritability, but South Koreans have been prosperous and well fed and North Koreans haven't, so the difference in height between them today is not genetic at all.

The problem with studies of adopted twins, as critics have noted, is that they usually involve children of fairly comfortable socioeconomic means. Even if the children are raised apart, they're still not at the poorest end of society, where good nutrition and a stable home life can be factors in their upbringing. They tend to go to fairly good schools. There's a risk, then, that these studies underestimate the role of environment across the true range of how people live.

But let's just assume for now that Robert Plomin and his twin studies are reliable, and that IQ tests are a fairly good measure of what we know as intelligence. Even if intelligence is at least half heritable among children raised under normal conditions, scientists like Plomin must also be able to point to the genes responsible for the effects they claim to see. They need to be able to explain, step by step, how they get from this twin correlation to the genes and then to the brain. To date, they haven't found anywhere near all the genes involved. In 2017 Plomin, along with a battalion of researchers in the Netherlands, Sweden, and the United States, published the results of a study of nearly eighty thousand people, which claimed to have found forty new genes linked to intelligence, bringing the known total to have such an effect to fifty-two. It was announced in the press as a breakthrough, but in reality these genes represented just a drop in the ocean. There are many, many more. "We're not talking about a handful of genes, we're talking about thousands of genes of very small effect," he admits. Half a century and millions of research dollars on, Arthur Jensen's prediction remains stubbornly unfulfilled.

Despite being unable to definitively isolate intelligence in human DNA, Plomin is proud of what he's achieved so far. He believes he is getting closer to an answer. "Do you know, two years ago we could explain 1 percent of the variance in intelligence with DNA? Now we can explain 10 percent! And it's only getting bigger. I would be amazed if at the end of the year we're not explaining 15 percent." Supposing he does, though, he would still be left with the challenge of finding a single mechanism, even one biological pathway, explaining how any of these genetic variations acts on the brain and leads to what we see as someone's general intelligence. We know, for instance, that X-linked mental retardation is a genetic condition, identifiable in a person's DNA, reliably leading to certain intellectual disabilities. There's a quantifiable link between the gene, inheritance, and cognition. But for everyday intelligence, scientists don't have anything clear like this.

Eric Turkheimer, a psychologist at the University of Virginia, believes they will never find it. "I've been around for a while," he tells me. "I've been in this field thirty years, and every single one of those thirty years, the biology people of one stripe or another have been saying, 'I know we're not there yet but in five years, as soon as this next piece of technology is nailed, as soon as we have brain scans, as soon as the genome project is completed . . .' It's always right around the corner. And the reason I don't believe it is because I don't believe that's the way genetic causation works."

Turkheimer compares intelligence to marriage. Psychologists know that if you have an identical twin who has been divorced, statistically you are more likely to be divorced yourself. There is no suggestion that there's such a thing as a gene for divorce, because people understand this to be a complex outcome, influenced by countless factors, including social and personality factors. "I think there are limits to how much we can understand something as complicated as divorce looking from the bottom up," Turkheimer says.

When it comes to intelligence, like most other complex traits, heritability depends crucially on context. He and his colleagues have seen, for instance, that in studies of people with the lowest socioeconomic status, environment explains almost all the variation researchers see in IQ, with genes accounting for practically *nothing*. Children who are the most socially and economically disadvantaged have been shown to lose IQ points over their summer holidays, while the most advantaged ones gain knowledge and skills over the same time period.

So for Turkheimer, it beggars belief that anyone should assume that the cognitive gaps psychologists now claim to see between racial groups in the United States could be biological. The effects of slavery and centuries of racism, in all its forms, are hard to quantify, but black Americans have undoubtedly suffered in ways that have left their marks on generations. "Millions of people were kidnapped and thrown in the bottom of boats and taken across the ocean, and a third of them died on the trip, and then thrown on plantations and enslaved for hundreds of years. And after that, treated with total discrimination. And now, *now* their IQs are a little lower? And we're saying it's in their *genes*? My feeling about that is, give me a break."

We know from the seeds-planted-in-two-pots example that you can't compare populations that live in different environments because the differences between them could easily be ascribed to nurture rather than nature. And there is no doubt that the social and economic circumstances of most black Americans remain significantly behind those of white Americans. The Institute for Women's Policy Research in Washington, DC, found that in 2017 the median weekly wage taken home by a white man working full-time was 33 percent higher than that taken home by a black man in a comparable job. It was almost 50 percent higher than that of a black woman. The wealth gap—"wealth" refers to assets accumulated over generations—is even starker. Research in 2017 showed that of lower- and middle-income households, white families have four times as much wealth as black families. According to the most recent data, two-thirds of black children live in single-parent families, compared with a quarter of children in "non-Hispanic" white families. Across the board, black Americans are significantly worse off, from the level of police brutality they suffer to the quality of healthcare and schooling they receive.

The logical consequence of insisting that IQ gaps between races are biologically determined is that nothing in human society can really be changed. In an age in which some like to believe that we have transcended the old rules of social inequality, when the playing field is supposed to be level, when women have the vote, when black Americans have civil rights, and colonialism is dead, they believe that biology is all that's left to explain the disparity that remains. Inequality, then, must be natural, the product of the survival of the fittest. Yet we still don't have the genetic evidence to prove any of this, says Turkheimer. All we have is the belief that the proof *will be* there somewhere in the genes.

"I don't see how we can get from where we are now to that kind of racial speculating that people like to do."

| | | | | | | | |

Turkheimer lives in Charlottesville, Virginia. In August 2017 the city became the infamous backdrop to a Unite the Right rally that drew together white nationalists, fascists, and neo-Nazis from across the United States, brandishing swastikas and Confederate flags. Their march eventually escalated into violence, culminating in the death of a peaceful antiracism protestor, Heather Heyer, who was struck when a car drove into her and others who had braved the threat of violence to challenge the message of those on the far-right. It was a day that was described by many as a wake-up call to America. The dream of a postracial society seemed more distant than ever.

"The synagogue we belong to is right next to the park where it all happened," recalls Turkheimer. A letter from his rabbi told a disturbing story. "It was Saturday morning, so there were services inside. They locked the doors while people marched up and down the street yelling, 'Burn it down.'"

In the following days, editors of the scientific journal *Nature* felt the need to run an editorial reaffirming that science cannot and should not be used to justify prejudice. It was a brief but remarkable statement, proving just how potent scientific racism was seen to have become. "This is not a new phenomenon," the editors remarked about Charlottesville. "But the recent worldwide rise of populist politics is again empowering disturbing opinions about gender and racial differences that seek to misuse science to reduce the status of both groups and individuals in a systematic way."

Turkheimer explains that the problem is not in the data, which is so far either unclear or unsupportive of racist interpretations of intelligence, but in the rampant speculation. If science could conclusively tell us that there was a biological difference in our DNA that made some groups smarter than others, there would be no more need for debate. "But my point is, I don't think we have that. And so what we're doing instead is speculating about our intuitions, speculating about people's dumb intuitions about Jews and blacks and whoever it is people like to speculate about."

If bad intuition is the problem, it's a problem we all have. Intelligence is just as multifaceted a cognitive trait as any other, but there's a widespread assumption that it is very heavily influenced by inherited natural ability. In reality, parents' IQ scores can only explain 15 percent of the variance in their own children, admits Plomin. Exceptionally smart parents are likely to have children a little less smart than themselves because of a phenomenon known as regression to the mean, which works to bring everyone in a population back closer to the average. Very bright children are likelier to emerge from parents in the middle of the intelligence range, where most people live. This was precisely the statistical fact that made eugenics impossible.

As the *Nature* editorial noted, "Every individual is a potential exception." Plomin himself is living evidence of just how possible it is for individuals to be unlike the rest of their family members. He was raised in a working-class family in Chicago, and neither of his parents went to college. "My sister is as different from me as you can be, in looks, in personality, and she was never interested in books. I would go to the library and get all these books, and she didn't go to university," he tells me.

Today, there remains little doubt that there is at least some heritable component to what we perceive to be an individual's ability to reason and solve problems, to process complex ideas and generally figure things out. But it's the degree of flexibility in there, in this hard-to-pin-down thing we call intelligence, that still eats away at some in the scientific community, not to mention many on the outside who are interested in the politics of it all. The debate is continually reduced to a simple question of nature and nurture, to biology and the environment.

But it's not such a simple equation. Even identical twins can sometimes show very different abilities. In Thomas Bouchard's twin studies, he came across a pair of brothers who were raised in wildly different environments. One grew up to be an uneducated manual laborer and the other was highly educated. An IQ difference of twenty-four points separated them. In the normal course of development, "a ten- or twelve-point difference between identical twins is not unusual," Turkheimer tells me.

"So how much flexibility might there be in the system?"

| | | | | | | | | |

In 1984 James Flynn, an intelligence researcher based at the University of Otago in New Zealand, raised a collective gasp in the scientific community when he announced that for nearly fifty years, since 1932, American IQs had been rising at a rate of about three points a decade. This finding could have been interpreted as people getting significantly smarter with each passing generation. In reality, as Flynn recognized, people had simply become much more skilled at taking IQ tests. This phenomenon is now known as the "Flynn effect."

Test takers were performing better not because they had evolved mental capacities beyond those of their grandparents but because their minds were being nurtured and sharpened now in ways they weren't before, by better and more education, more intellectually demanding jobs and hobbies. "The period in question shows the radical malleability of IQ during a time of normal environmental change," Flynn wrote at the end of his paper. Whatever link to intelligence is measured by IQ tests, it saw a benefit from the passage of time, from cultural change.

Comparing countries, Flynn saw similar effects everywhere. Different versions of IQ tests are used across the world, with varying results. But between 1951 and 1975 Japan saw an IQ gain of more than twenty points. In Britain it was almost eight points between 1938 and 1979. Countries that were fully modern before the testing period began tended to show more modest gains than those that underwent significant social and economic change. Kenya and Caribbean nations made particularly big leaps. Scandinavian countries, on the other hand, showed a peak and then arguably even a little decline. Flynn proved that there must be a great deal of flexibility in intelligence.

*Mankind Quarterly* editor Gerhard Meisenberg had told me that some countries are too cognitively challenged to prosper, that essentially they are poor because they are stupid. His only evidence was historical IQ test scores. Should anyone need it, the Flynn effect is some of the best proof yet that he is wrong. It shows that environment matters in IQ test results, even at a population level. In a paper published in *American Psychologist* in 2012,

Flynn, Turkheimer and other experts suggested that, at this rate, the apparent "IQ gap between developing and developed countries could close by the end of the 21st century." Flynn has shown that the IQ performance of African Americans has risen faster than that of white Americans in the same time period. Between 1972 and 2002 they gained between seven and ten IQ points on "non-Hispanic whites."

Another easily overlooked fact in the American black-white IQ gap debate is that very few African Americans are quite as ancestrally African as we might think. In 1976 sociologist Robert Stuckert used US Census data to estimate that as many as a 25 percent of people listed as white might reasonably have some African ancestors, and as many as 80 percent of black Americans were likely to have some non-African ancestors. In 2015 geneticists at Harvard University and researchers from the ancestry-testing company 23andMe investigated the backgrounds of more than five thousand people who self-identified as African Americans, and found that on average almost a quarter of their ancestry was European.

This should come as little surprise. The sexual exploitation of black women by their white owners was common during slavery. Thomas Jefferson is thought to have fathered children by a slave in his household, Sally Hemings, who was herself of mixed ancestry. And then, in the last century the United States has become ever more interwoven as the barriers between interracial relationships and marriage have been beaten down. Historically the "one-drop rule" meant that anyone with even the smallest degree of African ancestry—one drop of blood—was classified as black. Today, people who look at all "black," however complicated their heritage, are treated socially as black.

But if the biological portion of intelligence is rooted in a complex mixture of many thousands of genes, as biologists now agree it is, then it stands to reason that someone of mixed ancestry will have a mix of intelligence-linked genes from most if not all of their recent forebears. They wouldn't inherit genes from those with just one skin shade and not another. And if that's the case, and there are indeed innate, genetic racial differences in intelligence, then logic dictates that they should show up in people of mixed ancestry. If, as Rushton and Jensen implied, black people are biologically less smart, then shouldn't black Americans with higher proportions of white European ancestry have slightly higher IQs?

As far back as 1936, a study of exactly this kind was published by two American schoolteachers, Paul Witty and Martin Jenkins. They picked sixty-three of the highest-performing black children in the Chicago public school system, and compared their IQs with the proportion of white ancestry they were thought to have, according to information provided by their parents. The results revealed no gap at all. Having more white ancestry didn't raise a child's IQ. Indeed, the most remarkable student in the group, a girl with an IQ of 200, was reported to have no white ancestry whatsoever.

A similar study was carried out much later, in 1986, this time looking at black children who had been adopted into middle-class black and white families. Children who had one white parent and one black had around the same IQs as children with two black parents. What did make a difference in performance, though, was the family they were adopted into. The black and mixed-ancestry children adopted into white families had IQs thirteen points higher than those adopted into black families.

Andrew Colman, a psychologist originally from South Africa who now works at the University of Leicester in the UK, has interrogated the claims made by scientists who insist that intelligence gaps between races are real. He believes that research like this strongly indicates that environmental factors could well account for the entire black-white IQ gap in the United States. Even for those who claim to show contradictory evidence, he writes, "It is literally impossible to raise Black, White and mixed-race children in identical environments if racism itself is a significant environmental factor." Being in the same school or even in the same family means little if society as a whole sees you as substandard. Colman accuses researchers who cling to the idea that "negroes show some degree of genetic inferiority," as one once wrote, of a form of "self abuse."

It is interesting how the debate over racial differences in IQ takes on a different flavor in other countries. The United States seems to be a special case. In the United Kingdom, the group of sixteen-year-olds that achieves the lowest grades in IQ tests is white working-class boys, followed by white working-class girls. Yet scientists haven't leaped up to claim that low intelligence is rooted in whiteness. There's no evidence that being white in the UK is a socially disadvantaging factor either, so by this logic it must be their socioeconomic status that's the problem. In the decade to from 2006 to 2016, some of the greatest progress in educational attainment in the UK was seen among Bangladeshi, black African, and Chinese pupils. Girls have also historically tended to outperform boys, even though there is no average intelligence gap between the sexes. According to the founder of the Sutton Trust, which researches social mobility in the UK, it's clear that culture is at play here. There are social influences where class, ethnicity, and gender intersect, and they all affect achievement.

This is a point that even Robert Plomin—whose work has been described in *Nature* as "vintage genetic determinism"—concedes. He acknowledges that studying group differences in intelligence is fraught because it is impossible to control adequately for the environmental effects, adding that he doesn't see any value in studying racial differences in intelligence. "Based on what we know now, I don't see how you would do it. Certainly for black-white differences, people have tried for a very long time and I don't think we're any further along," he tells me. "We've had forty years of history of this, and it's just a lot of heat and no light."

Heritability does matter, he insists before I leave his office. This position is, after all, his bread and butter. But he surprises me by adding that if you measure individual differences and find some people are a lot better than others, "they all can achieve incredible levels of skill given culture."

| | | | | | | | | |

Sridhar Sivasubbu looked at me from behind his thick brush of a mustache.

I had asked him about his own community and what it meant to him. He told me that he came from a particular Tamil group in southern India known in ancient times for being warriors, whose members nowadays tend to join the military or the police force. "If you go back and read the history of my own community, largely people have been doing this for ages, so people just keep doing it. That's what they learned." As a scientist, he happened to be an outlier. I couldn't help but detect a hint of shame in his voice, or perhaps it was just regret, for the well-trodden path he chose not to take.

"People tend to follow what their ancestors did," he mused. In a country where roots go so far back in time that truth and myth are almost the same thing, connections to the past are not easily severed. They define how people live, forming the rigid framework of a sometimes precarious existence. "Let's say we go back a few hundred years. You still had a set of beliefs, you still had a set of skills, trades that would be passed in the community. So you learned it, and you used that for livelihood. There will be communities that will end up growing a certain type of crop, in a certain type of region," he explained. "It helped them survive. You also had the bonding, friends, family, neighbors. So it helped you in times of crisis."

But keeping close to a group has its costs. India's long history of community marriage has certainly left a mark in the genetic profiles of its inhabitants. These marks are most obvious when it comes to health. "A genetic disease tends to remain in the community," Sivasubbu told me. Studying particularly isolated communities has revealed stagnant pools. For the very smallest populations, reluctance to marry outside the group can be deadly. The Parsis, descended from Persians who moved to India more than a thousand years ago, suffer such high rates of particular forms of breast and prostate cancer, possibly inherited, that there have been fears the group may disappear altogether.

Sivasubbu has helped identify other Indian families with rare genetic illnesses, including one that left siblings covered in dried skin resembling the scales on a snake, and another that is a neurological disorder so severe that the parents wanted their children euthanized. These illnesses emerged for the simple reason that their communities were too close-knit. People would end up unwittingly married to someone with a common recent ancestor, and if they happened to share the same recessive genes for a rare disease, the disease was more likely to occur in their children. The same dynamic can be seen with Tay-Sachs, a genetic nerve disorder that is more common than average among Ashkenazi Jews and non-Jewish French Canadians living near the Saint Lawrence River. By isolating themselves, populations have formed tight bonds, but the smaller groups have also burdened themselves with higher genetic risks. Racial "purity" comes with a high price.

I asked Sivasubbu whether he had married within his own community or had chosen to abandon that practice, knowing the risks. He told me that, despite fully understanding the genetic problems, his culture was so important to him that he found a wife from within his group. So long as he followed certain rules around choosing someone not too closely related to him, he believed it was better to stay tight to his community than to leave it. "It was a very conscious decision, very conscious decision. It's going back to your roots, your beliefs," he said, smiling.

For me, as a person of Indian ancestry raised in London, between two cultures but somewhat detached from both, this was something I had never fully understood. In that moment it struck me just how powerful a thing culture can be. It can make us act against our better judgment, but it also anchors us in the world, in time and in place. Culture, and the safety and security it brings, can have such a profound impact on behavior over generations, that to outsiders the cause of the behavior may well appear to be genetic. It can seem to be woven into a person, unshakable, when in fact that same individual under other circumstances and raised in another place may behave completely differently.

When we see the effects of culture, we can't help but dream up biological mechanisms to explain it. The newly discovered travel diaries of Albert Einstein, written around 1922, have revealed that even he formed generalizations as he toured the world, despite being an antiracist humanitarian. He described the Chinese as an "industrious, filthy, obtuse people," adding, "It would be a pity if these Chinese supplant all other races." Perhaps he imagined, like the builders of the old human zoos, that our differences run all the way from our habits into our bones, that to know one Chinese person is to know them all.

Biologists Marcus Feldman at Stanford University and Sohini Ramachandran at Brown University have suggested that the "missing heritability" that scientists have for so long struggled to find in our DNA when it comes to intelligence and other complex traits may in fact be explained by the magic ingredient of culture. New scientific tools help us understand our genomes better, but they have only *reduced* the proportion of intelligence scientists now believe to be heritable. Feldman and Ramachandran ask the obvious: Why do scientists not look elsewhere for explanations?

In the same way that our parents pass on their genes to us, they also pass on their culture, their habits, their ways of thinking and doing things. And this can happen over generations. It is so sticky and persistent that it may well look biological to an observer. This is why measuring differences between groups, even over long periods of time, is laden with error. We are social beings, not just biological ones.

"The evolution of the Indian caste system is the perfect example of social determinism," an Indian biologist, Rama Shankar Singh, wrote in 2001. Studying the biological differences between castes, Singh started to understand it as a system defined not by evolved differences that made people better at different things, but as a set of barriers maintained by society for so long that they felt like they were in the blood. Lives and choices were constrained by the invisible forces of culture, and so everything remained stratified. Everyone kept their position for fear of stepping out. We don't change easily.

But we can change. As societies do shift and inequalities finally flatten, then we start to see our assumptions overturned. Stepping out of a rigid, unjust system can prove just how flexible we really are, just how far outside our genes our differences really may lie. In April 2018 a study was published that looked into the performance of Indian bureaucrats hired as result of affirmative action policies. The Indian Administrative Service, one of the largest and most powerful bureaucracies in the world, is also one of the toughest places to land a job. Of the 400,000 people who apply, 7,500 are invited to take a grueling exam, of which as few as 100 or so will be offered a position. Controversially, half of these vacancies are then reserved for marginalized castes, whose slightly lower scores would usually disqualify them.

A common assumption has been that, even though they help redress social inequality, these quotas must have an impact on standards. If people have to be given a leg up to get in, then they surely can't be as good? Some believe that those born into lower castes are innately incapable of doing these high-status jobs well anyway, regardless of their actual socioeconomic position. But when they investigated one particularly large sample project, American scholars Rikhil Bhavnani and Alexander Lee found no statistically significant difference in performance at all. "Improvements in diversity can be obtained without efficiency losses," they concluded. Caste made no impact.

Indeed, the lower-caste applicants who got through the usual way, without the quota, tended to perform somewhat better than average.

CHAPTER 10

# The Illusionists

*Down the rabbit hole of biological determinism*

"Once upon a time . . ."

At the turn of the millennium, excitement around the dizzying possibilities of genetic research was still high. People wondered whether gene therapy could someday cure cancer. Researchers imagined they would find genes for everything from being tall to being gay, whether we might even build designer babies by tinkering with our DNA. And two scientists working for the National Cancer Institute in the United States wrote a fairy tale.

Their protagonist was a well-meaning geneticist who one day begins to wonder why some people use chopsticks to eat their food and others don't. So of course, the hero does what all good experimentalists do: he rounds up several hundred students from his local university and asks them how often they each use chopsticks. Then he sensibly cross-references that data with their DNA and begins his hunt for a gene that shows some link between the two.

Lo and behold, he finds it!

"One of the markers, located right in the middle of a region previously linked to several behavioral traits, showed a huge correlation to chopstick use," the tale goes. He has discovered what he decides to call the "successful-use-of-selected-hand-instruments" gene, neatly abbreviated to SUSHI. The magic spell is cast. The experiment is successfully replicated, the scientist gets his paper published, and he lives happily ever after.

This might have been the end were it not for one fatal yet obvious flaw. It takes him as long as two years to hit upon the uncomfortable realization that his research contains a mistake. The SUSHI gene he thought he found just happened to occur in higher frequencies in Asian populations. So it wasn't the gene that made people better at using chopsticks; it was that people who used chopsticks for cultural reasons tended to share this one gene a little more often. He had fallen headlong into the trap of assuming that a link between chopsticks use and the gene was causal, when in fact it wasn't. The spell was lifted and the magic was gone.

Like all good fairy tales, there was a moral to this story. Although not everyone could see it.

| | | | | | | | | |

In 2005 the hype around genetics had begun to fizzle out, slowly replaced with a healthier skepticism. Scientists began to wonder whether our bodies might not be quite as straightforward as they had thought. And then along came a young geneticist at the University of Chicago in the United States with an extraordinary claim.

Bruce Lahn's work was a shot in the arm for those who had always hoped that genes could explain everything, for the biological determinists who believed we were anything but blank slates, that in fact much of what we are is decided on the day we're conceived. It was a claim so bold it implied that maybe even the course of history could be decided by something as tiny as one gene.

Lahn had originally emigrated from China to study at Harvard University, and soon became regarded as a cocky maverick who didn't follow instructions, who did things his own way. A while after arriving in the United States, he changed his name from Lan Tian to Bruce Lahn in honor of the legendary actor and martial arts expert Bruce Lee. The science journalist Michael Balter describes in a profile how once, when invited to go on a two-day hike with his colleagues, Lahn turned up with nothing but a jar of pickled eggs.

"He was kinda the whiz kid; he was kinda the darling," Balter recalls.

Lahn moved up the academic ladder at lightning speed. In 1999 he was named one of *MIT Technology Review*'s thirty innovators under thirty. Then in 2005 he published a pair of studies in the prestigious journal *Science* drawing a connection between a couple of genes and changes in human brain size. He and his colleagues stated that as recently as 5,800 years ago (just a heartbeat in evolutionary time), one genetic variant that was linked to the brain among other things had emerged and swept through populations because of evolution by natural selection. Their implication was that it bestowed some kind of survival advantage on our species, making our brains bigger and smarter. At the same time, he noted that this particular variant happened to be more common among people living in Europe, the Middle East, North Africa, and parts of East Asia but was curiously rare in Africa and South America. Lahn speculated that perhaps "the human brain is still undergoing rapid adaptive evolution"—although not for everyone in the same way.

His work was a sensation. What set pulses racing above all was his observation that the timing of the spread of this gene variant seemed to coincide with the rise of the world's earliest civilization, in ancient Mesopotamia, which saw the emergence of among the first sophisticated human cultures and written language. Lahn seemed to imply that the brains of different population groups might have evolved in different directions for the past five millennia, and that this may have caused the groups with this special genetic difference to become more sophisticated than others. In brief, that Europeans and Asians had benefited from a cognitive boost, while Africans had languished without it, perhaps were still languishing.

Racists ate it up and asked for second helpings. After all, here was hard scientific evidence that seemed to corroborate what all those nineteenth-century colonialists and twentieth-century contributors to the *Mankind Quarterly* had always claimed, that some nations were intellectually inferior to others. Their failures to be economically prosperous were rooted not in history, but in nature. "There will be plenty more results where these came from," predicted John Derbyshire, a right-wing commentator who wrote for the American conservative magazine *National Review*. Lahn also attracted support from the late Henry Harpending, a geneticist at the University of Utah and the coauthor of a controversial book arguing that biology could explain why Europeans conquered the Americas and also that European Jews had evolved to be smarter on average than everyone else.

But there were problems with Lahn's findings. Even if his gene variants did show up with different frequencies in certain populations, it didn't necessarily mean that they provided those who had them with a *cognitive* advantage. They were known to be linked to organs other than the brain, so if the variants were selected for, maybe this was nothing to do with intelligence. The hypothesis needed more evidence.

So, soon after the papers were published, the controversial Canadian psychologist John Philippe Rushton ran IQ tests on hundreds of people to see if possessing the gene variants really did make a difference. Try as he might (and we can reasonably assume that as the head of the Pioneer Fund at the time, he tried his hardest), he couldn't find any evidence they did. They neither increased head circumference nor general mental ability.

Before long, critics piled on across the board, undermining every one of Lahn's scientific and historical assertions. For a start, the gene variant he described as emerging 5,800 years ago could actually have appeared within a time range as wide as 500 to 14,100 years ago, so it may not have coincided with any major historical events. The respected University of Pennsylvania geneticist Sarah Tishkoff, who had been a coauthor on his papers, distanced herself from the suggestion that the gene variants in question might be linked to advances in human culture, as Lahn had suggested.

There were doubts, too, that Lahn's gene variants had seen any recent selection pressure at all. Tishkoff tells me that scientists today universally recognize intelligence as a highly complex trait, not only influenced by many genes but also likely to have evolved during the far longer portion of human history when we were all hunter-gatherers, until around ten thousand years ago. "There have been common selection pressures for intelligence," she explains. "People don't survive if they're not smart and able to communicate. There's no reason to think that there would be differential selection in different populations. That doesn't mean somebody won't find something someday. Maybe it's possible, but I don't think there's any evidence right now that supports those claims."

In the end, Lahn had no choice but to abandon this line of research. "It was pretty damaging, because a lot of illustrious researchers either couldn't replicate his original findings or did not come to the same conclusions," explains Michael Balter, who interviewed Lahn, his critics, and his supporters at the time. *Science*, the journal that published his papers, even came under attack for including the more speculative portions of his work in the first place.

Lahn was partly a victim of how science works these days. The big discoveries have been made, so researchers often have little choice but to drill down into small, specific areas within biology. To make a name, they need themselves and the world to believe that this little thing they're studying is significant. According to Martin Yuille, a molecular biologist at the University of Manchester, "If you're going to do an experiment you have to be reductionist. You have to look for one of the factors that is associated with a phenomenon, and you're tempted inevitably to try to think of that factor as being a cause, even though you know it is actually an association. So you're kind of driven to it.

"It is all too easy to exaggerate the role of the one variant of a gene that you might identify as associated with a trait. . . . But you need to be modest."

In this case, the world had seen the chopsticks fairy tale play out for real. In hindsight, it seems obvious that just because a genetic change in the brain may be more common in certain geographical populations than in others, that's no basis for claiming that it could be responsible for the economic or political fortunes of entire regions. By any measure, this was an irresponsible leap of faulty logic. But then Lahn was known to be cocky, to do things his own way.

Gerhard Meisenberg from the *Mankind Quarterly* made the same assumption when I interviewed him—that the innate abilities of a country's people are what define its success, even if we don't have any scientific evidence for it. It's an idea that has underpinned racist thought for centuries. It rests on the assumption that groups fall into ranks based on immutable biological features. It has the scent of the multiregional hypothesis, implying that nature has taken different tracks, that some of us are more "highly evolved" than others.

When I contact Lahn, now a professor of genetics at the University of Chicago, it has been more than a decade since his controversial papers were published. In 2009, undeterred by his embarrassing failure, he wrote a piece in another topflight journal, *Nature*, calling for the scientific community to be morally prepared for the possibility that they might find differences between populations, to embrace "group diversity" in the way that societies already cherish cultural diversity. He argued that "biological egalitarianism" won't be viable for much longer, implying that not all population groups are actually equal. He tells me that he's still "open to the possibility that there may be genetic differences in intelligence between modern populations, just like there may be genetic differences in other biological traits between modern populations such as bodily measures, pigmentation, disease susceptibility and dietary adaptation."

As I learn, his hypothesis hasn't changed. Lahn sticks firmly to the line that he is guided by science, wherever this may take him. "Before there is data, these are just possibilities," he explains. "My nose follows the scientific method and data, not politics. I am willing to let the chips of data fall where they may, as any self-respecting scientist should."

| | | | | | | | |

Barbara Katz Rothman, professor of sociology at the City University of New York, has written: "Genetics isn't just a science; it's a way of thinking. . . . In this way of thinking, the seed contains all it could be. It is pure potential."

For Eric Turkheimer, the assumption that propels race research today in all its various forms is a sign of this deterministic pathology. "There are people out there who think in a serious way that they're going to link up gene effects, the things you see in brain scans, the things you see on IQ tests," he tells me. They are looking for that elusive mechanism, that magic formula which will allow them to take the genomes of people from Europe, or Africa, or China or India, or anywhere else, and prove beyond a shadow of a doubt that one population group really is smarter than another. It's all there in our bodies just waiting to be discovered.

"It's a racist hypothesis," he adds.

The origins of this idea—that everything is in the genes—date back to the middle of the nineteenth century, when Gregor Mendel, an Augustinian friar in Brno, Moravia, then part of the Austrian Empire, became fascinated by plant hybrids. Working in the garden of his monastery, Mendel took seven strains of pea and bred them selectively until each one produced identical offspring every time. With these true-bred pea plants, he began to experiment, observing carefully to see what happened when different varieties were crossed. Nobody knew about genes at this point, and in fact Mendel's paper on the topic published in 1866 went largely unnoticed within his lifetime. But his experimental finding that traits such as color were being passed down the generations in certain patterns would form the linchpin of how geneticists in the following century thought about inheritance.

Once scientists understood that there were discrete packets of information in our cells that dictated how our bodies were built, and that we got these packets in roughly equal measure from each parent, the science of heredity finally took off. And it took almost no time for the political implications to be recognized. In 1905 the English biologist William Bateson, Mendel's principal popularizer, predicted that it "would soon provide power on a stupendous scale."

Mendelism became a creed, an approach to thinking about human biology which suggested that a large part is set in motion as soon as an egg is fertilized, and that things then work in fairly linear fashion. If you crossed one yellow-seeded pea plant with one green-seeded pea plant and you could predict which colors subsequent generations of pea plant would turn out to have, then it stood to reason that you might be able to predict how human children would look and behave based on the appearance and behavior of their parents.

Through a narrow Mendelian lens, almost everything is determined by our genes. Environment counts for relatively little because we are at heart the products of chemical compounds mixing together. We are inevitable mixtures of our ancestors. Just as Bateson foresaw, this idea became the cornerstone of eugenics, the belief that better people could be bred via the selection of better parents. "Mendelism and determinism, the view that heredity is destiny, they go together," says Gregory Radick, a historian and philosopher of science at the University of Leeds, who has studied Mendel and his legacy.

But Mendel's pea plant research had a problem. At the beginning of the twentieth century Mendel's paper became the subject of a ferocious debate, says Radick. "Should the Mendelian view be the big generalization around which you hang everything else? Or on the contrary, was it an interesting set of special cases?" When Mendel did his experiments, he deliberately bred his peas to be reliable in every generation. The aberrations, the random mutants, the messy spread of continuous variation you would normally see, were filtered out before he even began, so every generation bred as true as possible. Peas were either green or yellow. This allowed him to see a clear genetic signal through the noise, producing results that were far more perfect than nature would have provided.

Raphael Weldon, a British professor at the University of Oxford with an interest in applying statistics to biology, spotted this dilemma and began campaigning for scientists to recognize the importance of genetic and environmental backgrounds when thinking about inheritance. "What really bothered him about the emerging Mendelism was that it turned its back on what he regarded as the last twenty years of evidence from experimental embryology, whose message was that the effects a tissue has on a body depend radically on what it's interacting with, on what's around it," explains Radick. Weldon's message was that variation matters, and that it is profoundly affected by context, be it neighboring genes or the quality of air a person breathes. Everything can influence the direction of development, making nurture not some kind of afterthought tacked onto nature, but something embedded deep in our bodies. "Weldon was unusually skeptical."

To prove his point, Weldon demonstrated how ordinary pea breeders couldn't come up with the same perfectly uniform peas that Mendel had.

Real peas are a multitude of colors between yellow and green. In the same way that our eyes aren't simply brown or blue or green, but a million different shades. Or that if a woman has a "gene for breast cancer," it doesn't mean she will necessarily develop the disease. Or that a queen bee isn't born a queen; she is just another worker bee until she eats enough royal jelly. Comparing Mendel's peas to the real world, then, is like comparing a soap opera to real life. There is truth in there, but reality is a lot more complicated. Genes aren't Lego bricks or simple instruction manuals; they are interactive. They are enmeshed in a network of other genes, their immediate surroundings and the wider world, this ever-changing network producing a unique individual.

Sadly for Weldon, the ferocious debate for the soul of genetics ended prematurely in 1906 when he died of pneumonia, aged just forty-six. His manuscript remained unfinished and unpublished. Without as much resistance as before, Mendel's ideas went on to become incorporated into biology textbooks, becoming the bedrock of modern genetics. Although Weldon's ideas have since slowly returned into scientific thinking, there remains a strain of genetic determinism in both the scientific and public imaginations. Harvard biologist Richard Lewontin called it the "central dogma of molecular genetics." It is a belief that it's really all there in our genes.

In 2015 sociologists Carson Byrd at the University of Louisville and Victor Ray at the University of Tennessee investigated the belief of white Americans in genetic determinism. Studying responses to the General Social Survey, which is carried out every two years to provide a snapshot of public attitudes, Byrd tells me they found that "whites see racial difference in more biologically deterministic terms for blacks." Yet they tend to view their own behavior as more socially determined. For instance, if a black person happens to be less smart than average, the whites attribute this to the black person's having been born this way, whereas a white person's smartness or lack thereof is seen more as a product of outside factors such as schooling and hard work. "So they give people a bit more leeway if they're white," he explains.

Also interesting to Byrd was that even though the General Social Survey found that white conservatives were a little more biologically deterministic than white liberals, people with this view on both sides of the political spectrum shared the belief that policy measures such as affirmative action are needed to improve the lot of black Americans. There's a slippery slope here, he warns. "The slipperiness is that they believe that because it's genetic, they can't help themselves, that it's innate, that they're going to be in a worse social position because of their race." In other words, they want society to help, not because they believe we're all equal underneath, but because they believe we're not.

"Before it was something in the 'blood' and now it's in our genes," Byrd tells me. What has remained the same over the centuries are the racial stereotypes of black Americans. Rather than black disadvantage being seen as social or structural in origin, which it is, it's conveniently rendered in the new scientific language of genetics. "A lot of people have become enamored with the science . . . the mystique of things that could be embedded within our genes."

Stephan Palmié, an anthropology professor at the University of Chicago, has argued that even now, "Much genomic research proceeds from assumptions it culls from ostensibly 'scientific' constructions of the past . . . and eventually restates them in the form of tabulations of allele frequencies" (alleles are different forms of the same gene). Nineteenth-century ideas about race that have gone out of fashion take on an almost magical quality when they're freshly rewritten in the language of modern genetics. Today there is technical jargon, charts and numbers. Suddenly the ideas seem shinier and more legitimate than they did a moment ago. Suggest to anyone that the entire course of human history might have been decided by a single gene variant and they'll probably laugh. But that's exactly what Bruce Lahn did suggest in the pages of one of the most important journals in the world. For a moment, it felt possible because it was new science.

The belief that races have natural genetic propensities runs deep. One modern stereotype is that of superior Asian cognitive ability. Race researchers, including Richard Lynn and John Philippe Rushton, have looked at academic test results in the United States and speculated that the smartest people in the world must be the Chinese, the Japanese, and other East Asians. When the intelligence researcher James Flynn investigated this claim, publishing his findings in 1991, he found that in fact East Asians had the same average IQ as white Americans. Remarkably, though, Asian Americans still tended to score significantly higher than average on the SAT college admission tests. They were also more likely than average to end up in professional, managerial, and technical jobs. The edge they had was therefore a cultural one: their upbringing had endowed them with more supportive parents or maybe a stronger work ethic. They just tended to work harder than others.

To anyone who has grown up as an ethnic minority anywhere, especially those of us who were told that we have to work twice as hard to achieve the same as white people, this will come as small surprise. Among middle-class Indians living in the United Kingdom (the group my parents belong to), the weight of cultural pressure has generally been on children to become physicians, pharmacists, lawyers, and accountants. These are professions that tend to be well respected and well paid, with no shortage of job opportunities and straightforward entry once you have the right qualifications. They are reliable routes into middle-class society. Medicine carries such an immense prestige bias among immigrants and children of immigrants that, according to the most recent data gathered by the British Medical Association, around a quarter of all British physicians are South Asian or have South Asian heritage. In the United States, the American Association of Physicians of Indian Origin boasts eighty thousand members. This is not because Indians make better doctors but because culture acts as an invisible funnel, not dissimilar to the way women get channeled into caring professions such as nursing, because this is what society expects. Culture molds people, even subconsciously, for certain lives and careers.

It is also interesting how these stereotypes can change over time. Asian Americans are today considered a model minority. We forget that more than a century ago, unlike today, European race scientists saw Asians as biologically inferior, situated somewhere between themselves and the lowest-status races. In 1882 the United States passed the Chinese Exclusion Act to ban Chinese immigrant laborers because they were seen as undesirable citizens. Now that Japan has been highly prosperous for decades and India, China, and South Korea are fast on the rise with their own wealthy elites, the stereotypes have shifted the other way. As people and nations prosper, racial prejudices find new targets. Just as they always have.

| | | | | | | | |

"Think about what happened to all the old racial stereotypes," Eric Turkheimer challenges me.

"A hundred years ago, people were quite convinced that Greek people had low IQs. You know, people from southern Europe? Whatever happened to that? Did somebody do a big scientific study and check those Greek genes? No, nobody ever did that. It's just that time went on, Greek people overcame the disadvantages they faced a hundred years ago, and now they're fine and nobody thinks about it anymore. And that's the way these things proceed. All we can do is wait for the world to change and what seemed like hardwired differences melt away and human flexibility just overwhelms it."

But the waiting is hard. And as we wait, it remains all too easy for researchers to allow their assumptions about the world to muddy the lens through which they study it, and for the research they then produce to impact or reinforce racial stereotypes.

In 2011, Satoshi Kanazawa, of the Department of Management at the London School of Economics, who writes widely on evolutionary psychology, speculated that black women are considered physically less attractive than women of other races. "What accounts for the markedly lower average level of physical attractiveness among black women?" he blogged in *Psychology Today*, racking his brain. "Black women are on average much heavier than nonblack women. . . . However, this is not the reason black women are less physically attractive than nonblack women. Black women have lower average level of physical attractiveness net of BMI [body mass index]. Nor can the race difference in intelligence (and the positive association between intelligence and physical attractiveness) account for the race difference in physical attractiveness among women," he continued, in the manner of a drunk uncle.

At a stroke, Kanazawa took it as a scientific given that black women are both less attractive, which is obviously a value judgement, and innately less intelligent, which is unproven. Presenting these two offensive statements unchallenged, he landed on the speculative conclusion that their unattractiveness, as he had now established it, might have something to do with having different "levels of testosterone" from other women. Kanazawa, whose published work has since looked at intelligence and homosexuality, among other things, had his online post promptly pulled down under the weight of public and academic outrage. The London School of Economics banned him from publishing any more non-peer-reviewed articles or blog posts for a year.

But how did it get published at all? When Kanazawa invoked race as a factor in why he perceived some women to be more attractive than others, he was performing a sleight of hand. He was diverting attention away from the underlying question of where his assumptions came from, or why he was asking this specific question in the first place. In so doing he shone the spotlight straight onto his concluding statements. As soon as we, the audience, accepted his assumptions, his racist question could be transformed into a scientific one. It could seem almost legitimate rather than simple offensive.

Diverted, the publisher of his work failed to notice that his hypothesis was dripping with prejudice. It had no rigor to it at all.

The American sociologist Karen Fields has compared the use of the idea of race, as in this example, to witchcraft—using the word "racecraft." Race is commonly described by scientists, politicians, and race scholars as a social construct, as having no basis in biology. It's as biologically real as witches on broomsticks. And yet, Fields writes that she sees the same "circular reasoning, prevalence of confirming rituals, barriers to disconfirming factual evidence, self-fulfilling prophecies" among scientists that is common in folk belief and superstition. It almost doesn't matter what anyone says because race feels as real to us as magic feels real to those who believe in it. It has been made real by overuse.

When Bruce Lahn, just four years after he was forced to retreat from his flawed research on intelligence genes, asked the scientific community to embrace "group diversity" (that is, differences among groups), exactly what was he asking them to embrace? As he admitted to me himself, we don't yet have data on the differences between populations beyond superficial ones, and even these superficial variations show enormous overlap. The chips haven't yet fallen. His plea is not for us to accept the science we have, but to accept in advance something we don't yet know. He is assuming that data will eventually confirm what he suspects—that there are cognitive differences between groups—and is telling us to take his word for it. But how scientific is that? How close are his assertions to being simply statements of belief?

"I do science as if the truth mattered and your feelings about it didn't," Satoshi Kanazawa states on his personal website, which lacks any sign of remorse for his paper on black women. In 2018 he and a colleague at Westminster International University in Tashkent, Uzbekistan, published a paper in the *Journal of Biosocial Science*, produced by the reputable Cambridge University Press, asking why societies with "higher average cognitive ability" have lower income inequality. Again, he started with the assumption that scientists believe that populations have different cognitive abilities—again, unproven. Again, the editors failed to notice.

Among the very few researchers to have written on links between race, intelligence, and the wealth of nations are Gerhard Meisenberg, Richard Lynn, and Tatu Vanhanen, all intimately associated with the *Mankind Quarterly*. In a joint publication they have claimed that Africans have an average IQ of about 70. But when a Dutch psychologist, Jelte Wicherts, investigated this figure, he found they could have arrived at it only by deliberately excluding the vast majority of data that actually shows African IQs to be higher. "Lynn and Meisenberg's unsystematic methods are questionable and their results untrustworthy," he concluded. Even so, in his own work Kanazawa cites heavily from their work.

It's a problem that continues outside ivory towers and fringe journals. In 2013, Jason Richwine, a public policy researcher at a powerful conservative think tank, the Heritage Foundation in Washington, DC, was forced to resign after it was revealed that he had written a doctoral thesis while at Harvard University in which he claimed that the average IQ of immigrants into the United States was lower than that of white Americans. Richwine expressed the possibility that Hispanics might never "reach IQ parity with whites," ignoring that nobody considers "Hispanics" a single genetic population group, since they have such diverse ancestries. For instance, most Argentines are of European ancestry, just like white Americans are. Richwine created the illusion that Hispanics are a biological race.

From this, he followed up that immigration policy should focus on attracting more intelligent people. Upon joining the think tank, he also happened to cowrite a study suggesting that legalizing illegal immigrants, most of whom are Mexican and Central American, would result in an economic loss of trillions of dollars.

In January 2018, during a closed meeting on new immigration proposals held in the Oval Office, President Donald Trump reportedly asked lawmakers, in reference to immigrants from Haiti, El Salvador, and Africa, "Why do we want all these people from shithole countries coming here?" By the summer of 2018, a crackdown on illegal immigrants by the Trump administration resulted in thousands of young children being separated from their parents, wailing in distress and reportedly held in cages. He is believed to have stated in his January comments that the United States should be welcoming more immigrants from countries such as Norway.

| | | | | | | | |

The notion that there are essential differences between population groups, that genetically "shit" people come from "shithole countries," may be an old one, but the science of inheritance helped propel these racially charged assumptions into modern intellectual thought. The concept of genetic determinism has made some succumb to the illusion that every one of us has a racial destiny.

In reality, as science has advanced, it has only become clearer that things are more complicated. "We can't sidestep the fundamental problem that biological systems are *systems*; they are collections of organizations of matter that interact with each other and each of their environments," explains molecular biologist Martin Yuille.

He offers the example of diabetes, a disease believed to run in families. In the United Kingdom, it has been estimated that those of South Asian ancestry can be up to six times as likely as other ethnic groups to receive a type 2 diabetes diagnosis. But even if some people may be genetically slightly more predisposed to a disease, this isn't the same as actually receiving a diagnosis. Type 2 diabetes is well known to be heavily associated with lifestyle, such as diet and exercise, as well as age. Waist size is one of the most reliable correlations of all, which is why diabetes is becoming common among the middle classes of South Asia. Diets have always been rich in sugar and fat, but now people there are wealthier and more sedentary, leading to obesity. If diabetes were purely genetic, the world wouldn't be seeing a diabetes epidemic at this moment and not before. "Just to say that it runs in your family is ludicrous," says Yuille. "And it is fatalistic because it inhibits you from participating in activities that reduce risk."

This is what it means to be human, says Gregory Radick, "to know what it is to be a person and a body, to understand yourself along with everything else around you as this product of interactions between what you've inherited and your surroundings." Then you can begin to see that you "haven't been dealt a hand that you just have to accept, but it's within our gift to change these things and to make improvements. When you change the context, you can potentially change the effects."

Another example of a condition that scientists believed is heritable is schizophrenia, a mental disorder for which people of black Caribbean ancestry living in the United Kingdom receive disproportionately more diagnoses than white people, to the point where it has even been described as a "black disease." In recent years there have been feverish hunts for the genes thought to be responsible. In 2014 an enormous study involving more than 37,000 cases finally did find a number of genetic regions that may be associated with schizophrenia. But it turned out that the presence of even the most likely of these variants elevated the risk of suffering from schizophrenia by just a quarter of 1 percent over the risk in the population as a whole. One particular gene variant turned up in 27 percent of patients but also in around 22 percent of healthy subjects.

If schizophrenia is inherited, then its inheritance clearly can't be a straightforward equation. Indeed, environmental risk factors, including living in an urban environment and being an immigrant, have already been shown to be at least as important to being diagnosed as any genetic links found so far. One study published in *Schizophrenia Bulletin* in 2012 found that patients with psychosis were almost three times as likely to have been exposed to adversity as children. That's not to say the disorder doesn't have a genetic component, but it does demonstrate that it can't be quantified by looking at genes in isolation. If there are racial differences in diagnoses, it may be that life experiences, perhaps even the negative experiences resulting from racial discrimination, tip some people over the edge while rescuing others. This is without even considering that schizophrenia diagnosis itself is known to be notoriously subjective.

And if race is a factor, it's interesting to contrast the characterization of schizophrenia as a "black disease" with an observation by the Nazi scientist Otmar von Verschuer who worked at the Kaiser Wilhelm Institute of Anthropology, Human Heredity, and Eugenics in Germany. A year before the outbreak of World War II he wrote, "Schizophrenia is strikingly more frequent among Jews. According to statistics from Polish insane asylums, among insane Jews schizophrenia is twice as common as among insane Poles." Then he made a leap, twisting a medical observation into a racial generalization: "Since it is a matter of a hereditary disease . . . the more frequent occurrence of the disease in Jews must be viewed as a racial characteristic."

At that moment in time in that particular place, then, it wasn't a black disease; it was a Jewish one.

CHAPTER 11

# Black Pills

## *Why racialized medicine doesn't work*

"An Englishman Tastes the Sweat of an African" reads the caption of the black and white engraving.

Dated 1725, it documents everyday life in the slave trading post of Calabar, West Africa. Part of a work that later became a kind of instructional guide for European seafarers, it shows a man being examined before he's sold into slavery and shipped to the New World. There is an obvious inhumanity to it, the flavor of a cattle market. But it's made unnerving also by showing a strange act that is the engraving's central focus. The tall, well-built African slave dressed only in a loincloth crouches on his knees so that a skinny-legged, fully clothed Englishman with a sword hanging by his side can reach his face with an outstretched tongue to *lick his chin*.

In the following engraving, a ship sets off with its cargo of slaves, as families left behind mourn their loss, heads in their hands. The fatal transaction is complete. But there's still the memory of that odd lick. People have wondered just what the Englishman was doing. The accompanying description says he was confirming the slave's age and checking that he wasn't sick, although it's not immediately clear how a lick would achieve this. We can only assume there was some method to the technique.

Centuries later a young economist found the picture and came up with an alternative explanation. The licking, Roland Fryer at Harvard University has suggested, may have been to gauge the saltiness of the African man's body, because being a little brinier might better equip him to handle the long sea voyage to the New World. It's an idea inspired by a scientific theory that black Americans, mostly the descendants of slaves, process salt differently from white Americans. The slave ships that transported people from places like Calabar to the New World would have seen immense loss of life along the way as a result of fluid loss caused by dehydration, vomiting and diarrhea, the theory goes. The proportion who naturally retained more salt would have fared better, producing a genetic bottleneck among the slaves who survived to the end.

This process of human evolution on a rapid scale, acting only on black slaves, left them fundamentally different at the other end. Those who reached America were the saltiest.

Fast-forward to the present day. In the twenty-first century, this historical theory has been commandeered to explain why black Americans today suffer persistently high blood pressure, commonly known as hypertension, at higher rates than other groups in the country. Hypertension is made worse by eating too much salt, but if black Americans naturally happen to retain more salt because of the legacy of the slave trade, then the theory goes that this might be the reason they suffer hypertension, rather than because of their diets. Their bodies, some scientists and doctors argue, simply aren't the same.

| | | | | | | | | |

Hypertension is a simple thing to measure and it's a widespread problem all over the world. My mother happens to have it. In and of itself, it doesn't cause her any problems, but as the doctor reminds her, she needs to watch her sodium intake to bring her blood pressure down because high blood pressure increases her risk of heart disease and stroke. When I visit her for lunch I can't help but notice how liberal she is with the salt cellar. She doesn't need to remind me that it's a source of shame to serve Indian food underseasoned.

"Hypertension is probably the oldest chronic disease we know about," says Richard Cooper, a seventy-three-year-old public health researcher at Loyola University Medical School in Chicago. Cooper has spent decades investigating blood pressure, stirred by his days as a medical student in eastern Arkansas when he saw patients dying of strokes while still only in middle age. Part of the problem, he has found, is that it's a strangely nebulous disease. "There are some people who argue that hypertension is not a disease, it's a condition, that it's a state, like anxiety," he says. "We don't really know initiating cause." It isn't rooted in any one organ of the body; you can't run a scan or a biopsy for it. "It just emerges out of the mists."

Yet it's a major killer. The World Health Organization estimates that raised blood pressure accounts for nearly 13 percent of all deaths worldwide.

Despite how common hypertension is, only in around the 1940s or 1950s did American physicians begin to notice more black patients than usual coming to them with hypertension. In the United States, studies suggest that hypertension is almost twice as common in black Americans as in other groups. If you search the UK National Health Service website for the factors associated with high blood pressure, alongside salt you'll see lack of exercise, too much alcohol, smoking, older age, and one more: being of African or Caribbean descent. Hypertension is thought to be so powerfully correlated to blackness that UK clinical guidelines even recommend different drugs for black people and white people under the age of fifty-five.

Preventable heart disease and stroke resulting from high blood pressure are two to three times as likely to kill a black American as a white American. What is stranger is that this mirrors death rates from other causes, too. The life expectancy of a black person born in the United States today is three and a half years lower than that of a white person. Almost every major cause of death and disability hits blacks harder if they live in the United States; even infant mortality among the black population shows the same elevated rates. So in a country where black people die at disproportionate rates anyway, the mysterious quantity known as hypertension has long been a vehicle for speculation about racial differences.

Cooper saw it for himself growing up in Little Rock, Arkansas, and when he was at medical school in the 1960s. "There were patients who were getting a transfusion and would say, "You're not going to give me black blood are you?" You were constantly confronted with that," he recalls. At that time, wards weren't integrated and black Americans had access only to lower standards of healthcare. Growing up in Arkansas, he says, was like living under apartheid.

The notion of black exceptionalism runs right through the history of American medicine, Cooper adds. "There were very raw and basic racist ideas that were the norm in the medical school." One common belief was that some diseases, such as tuberculosis and syphilis, manifested differently in blacks. Another was that even if black and white bodies were similar physiologically, black bodies had less value. In the Tuskegee experiment of 1932, the United States Public Health Service teamed up with researchers at the Tuskegee Institute in Alabama (at the time a college for black people) and deliberately denied black patients antibiotics they knew could cure them so as to track the effects of syphilis. The men were observed until they died, their internal organs slowly ravaged. Only after ethical concerns were aired in the press did the study end, forty years later.

Hypertension is one of the conditions to have survived this era with its racial stereotypes intact. American doctors have for decades wondered whether the gaps they see could be due to some intrinsic difference between races. They used to ask whether hypertension might even be a uniquely different illness in black people, connected somehow to skin pigmentation or testosterone levels, or to the heat and humidity of Africa.

But as time passed and population studies were done, it turned out that people living in Africa, especially rural Africans, have the lowest levels of hypertension in the world. "The people are skinny, they don't eat much salt, they're very active. There is no way that blood pressures cannot be low. It's just not possible. They don't have diabetes and they don't have hypertension," says Cooper, who has carried out blood pressure studies on tens of thousands of people across Africa, Europe, and the Americas. People in Nigeria and Ghana in West Africa, from where most black Americans can trace their ancestry, are known to have far lower blood pressures than in other countries.

In fact, Cooper adds, topping the hypertension charts are Finland, Germany, and Russia. "They have terrific hypertension." White North Americans and Canadians, meanwhile, tend to have lower levels than Europeans, including those in England, Spain, and Italy. Hypertension, then, isn't a global problem for those with black skin; it's a local one. We know that black *Americans* certainly do have higher rates of hypertension on average than white *Americans*, and the same appears to be the case in Britain with black and white Britons.

For US physicians, the question has been, "Why do black Americans have higher rates of hypertension?" In the 1980s, attempting to neatly square the circle of low hypertension in black Africans with high hypertension in black Americans, a doctor named Clarence Grim came up with what became known as the "slavery hypertension hypothesis"—the theory, later championed by the Harvard economist Roland Fryer, that black Americans were naturally predisposed to retain more salt because of a rapid process of natural selection on the slave ships that brought their ancestors to the New World.

It was an evocative story, giving the tragic brutality of slavery an extra poignancy. Sensitivity to salt, which had helped some through the brutal journey across the Atlantic, landed their unfortunate descendants in the twentieth century with the fatal scourge of hypertension. Western diets had damned them, and there was nothing they could do. The media loved it. Fans of the theory included Oprah Winfrey and the resident health expert on her talk show, Doctor Oz. With Fryer's apparent historical evidence supporting Grim's hypothesis, combined with the picture of the slave being licked—it all seemed to be tied up in a bow.

Not everyone was taken in by the tale, though. Biologists raised eyebrows at the suggestion that evolutionary change could happen over such a short time scale. The historian Philip Curtin, an expert on the African slave trade, argued that dehydration and salt depletion were not significant causes of death on slave ships. Neither, he noted, had there ever been a shortage of salt in West Africa to make people there particularly more salt retaining. If anything, Curtin concluded, the historical evidence ran counter to the hypothesis rather than supporting it. Richard Cooper adds that salt sensitivity, if it is seen to be higher in black Americans, is likely to be a product of being primed over a lifetime for all the factors that give them hypertension in the first place, particularly poor diet. This is why other demographic groups who have higher hypertension, including men and the elderly, are also more sensitive to salt.

Even so, there was a widespread expectation that harder proof would one day be found, most likely in our genes. If anything could settle the debate once and for all, it would be the glittering new science of genomics.

| | | | | | | | |

In 2009 researchers thought they had finally discovered the evidence they were looking for. A team led by scientists at the National Human Genome Research Institute in Bethesda, Maryland, took DNA samples from around a thousand people and discovered five genetic variants linked to blood pressure in black Americans. The effects were admittedly modest, but they seemed promising. In that moment it felt as though there really might be a whole host of tangible genetic differences between races that would help science get to the root of disproportionate ill health in black Americans.

The search for "black genes" already had a precedent in the research surrounding sickle cell anemia, a serious blood condition more prevalent in people who have ancestry in malaria-afflicted regions such as West Africa.

The sickle cell gene is known to provide some resistance to malaria, which can otherwise be fatal. This is a watertight evolutionary explanation for why such a debilitating illness has persisted. But sickle cell disease also exists outside Africa, including in Saudi Arabia and India, which means it can be found in people of all different skin shades. As the UK's National Institute for Health and Care Excellence states, the sickle cell gene is actually found in *all* ethnic groups. Within the African continent, it isn't seen in high rates everywhere; for example, in South Africa malaria is less of a problem and the incidence of the sickle cell is therefore lower.

In the United States these nuances were lost purely for demographic reasons. Many white Americans tend to be of European extraction, where sickle cell anemia is rare, and black Americans tend to have West African roots, where it's more common, so it came to be seen as a "black disease." Once viewed this way, it reinforced existing assumptions about essential differences between blacks and whites. Two independent facts began to align in people's minds. First, that there may be different genes determining health according to race. Second, if black people suffer illness and death at higher rates than white people, could this then be genetic?

Hypertension, however, turned out to be not so straightforward. In 2012 another team of researchers, Clarence Grim among them, tried to replicate the 2009 study, this time with more than twice as many people. They failed. They just couldn't see the same correlations.

As scientists struggled to find the genetic evidence they thought must be out there, one team led by a researcher at the Harvard School of Public Health decided to look at factors other than race that might correlate with high blood pressure. They discovered that level of education, which often correlates with income level and social class, was a far better predictor of someone's having hypertension than the person's percentage of African ancestry. Each year of education was associated with an additional .5 millimeter decrease of mercury in blood pressure readings. A year later, a study in Cuba showed that being black or white there made no difference to average blood pressure or hypertension. Others pointed out that living in an urban environment was strongly associated with blood pressure rises, as was being an immigrant or adopting a westernized lifestyle. Richard Cooper suggests, too, that the effects of chronic exposure to discrimination could well account for some of the black-white differences in blood pressure in the few countries where they're seen.

Concerns were raised that the nongenetic influences on blood pressure were being oddly neglected by most medical researchers, even though this looked like an obvious medical gold mine of explanations for disparities in health.

"It has been very hard to find genetic factors that affect salt sensitivity. What has been found is not more common in blacks than whites," concludes Cooper. "There is no evidence of any significant selection across the African diaspora." Despite heavy resources poured into finding a gene, researchers have still found no association or mechanism that can fully account for higher hypertension in black Americans. The weight of evidence so far points elsewhere. If hypertension is more common among certain groups in certain countries, it probably isn't because of something in the genes. Most of it is likely to be down to the same reasons that my mother has hypertension—diet, stress, and lifestyle.

"Diet is the underlying cause of hypertension," Cooper insists. In Finland, for example, with among the highest rates of hypertension in the world, diets have traditionally been low in fruits and vegetables and high in fatty meat and salt. Food in the American South, traditionally associated with black Americans, is similarly rich in salt and fats. Cheap processed foods also have more added salt. Studies have shown that reducing salt intake by just 6 grams a day could prevent 2.5 million deaths annually from stroke and coronary heart disease worldwide. For scientists seeking an easy explanation for differences in hypertension rates across different groups, there's one available in every kitchen: the food people eat. Part of the reason that diet is understudied is that it isn't easy to measure with precision. People don't always know what's in the food they're eating, and they rarely report accurately what they do eat. It's easier to measure simple variables such as race, age, and sex.

But there's more to the lure of the slavery hypertension hypothesis than data collection, says Cooper. "It's a very useful window into people's thinking. Below the surface, there is this very strong prior bias to believe in a mechanism like that. Racialized thinking is such a deep part, just like gendered thinking is such a deep part, of our psychology that we can't just by conscious effort free ourselves from it completely. It keeps popping up in ways when we're unprepared or not vigilant."

People wanted so much for the story to be true, to be able to link the trauma of slavery to the trauma of black American deaths today, that they couldn't look past it to more mundane explanations. Hypertension, Cooper says, is a case of science being retrofitted to accommodate race. The data, the theories, the facts themselves, are rotated and warped until they fit into a racial framework we can relate to. This is the power of race. It is the power to twist science to its own ends.

| | | | | | | | |

"Typically in my field, we don't just collect data and then show the data that we collected," I'm told by Jay Kaufman, an epidemiologist and statistician based at McGill University in Canada. Data in its most raw form is rarely useful. Almost always, it has to be packaged, interpreted in some way to make sense.

He gives me a simple example. In 1970 the death rate in the city of Miami, Florida, was 8.92 per thousand people. In Alaska at the northwest tip of the United States, that same year it was only 2.67. This is the unvarnished truth. If you happen to live in Miami, you are more likely to die than if you live in Alaska. But then, if you happen to live in Miami, you're also more likely to be retired. In 1970 the city, with a total population around 60 percent of the whole of Alaska, had more than 92,000 people over the age of 65, while Alaska had only around 2,000. If you're between 15 and 24 in either of these places, the death rates are around the same. So statisticians adjust death rates by age to give more useful comparisons. Miami and Alaska then turn out to be about equally lethal as each other.

Adjustments like these are an everyday part of data analysis. Very little of what the public is fed in terms of statistics is unadjusted, and that includes health and medical data. "The logic of these statistical adjustments is based on ideas from trials," explains Kaufman. In experiments for new treatments, the gold standard is the double-blind randomized clinical trial, in which patients who are roughly the same in every other respect are randomly selected to receive either the treatment or a placebo but neither they nor the researchers know who has got what. Researchers can then be sure that the effects they see are because of the drug and not influenced by expectations.

In real life, though, it's not always possible or ethical to carry out randomized trials, which is why adjustments are made after the fact, artificially removing the effects of variables such as age and weight. When they do this, says Kaufman, researchers create what are in essence imaginary worlds where these things no longer matter. They are worlds made of manipulated data, sending a clear signal through the noise of reality. "This is sort of my niche within this field, the use of statistics and comparisons." He has years of experience in this, the darkest art of statistics. Kaufman is a traveler through imaginary worlds.

"Once we start adjusting and describing some imaginary world, then the question is: What is that for? Why are we making an imaginary world? Which imaginary world are we going to make? What are we trying to learn from this imaginary world that we can't learn from the real world?" When it comes to data about racial difference, the logic of adjustment is that if all the social and environmental aspects of racial difference can be removed and the data still shows a gap between groups, the cause of the gap must be biological.

In one study published in the *Journal of Allergy and Clinical Immunology* in July 2017 Kaufman spotted an article by a large team of American medical doctors who claimed that the airways of black people become more inflamed than those of white people when they have asthma. In the United States, it's well known that black Americans carry the heavier burden of asthma. They're almost three times as likely to die from asthma-related causes than non-Hispanic white Americans. Black children are four times as likely as white to be admitted to hospital for asthma. It's also well known that asthma is affected by the environment, including smoking, air pollution from busy roads and factories, and living among cockroaches, dust mites, and mold. Yet this study claimed that there was also something intrinsic to their bodies that made black asthma patients suffer more severely. The authors stated that black people might even need their own therapies.

Kaufman decided to pick through their figures.

The original data the researchers had on airway inflammation showed no significant differences between white and black patients. So they decided to control for factors including lung function and degree of control of the disease, as well as body mass index, age, and gender, until finally they came up with adjusted data that showed that there *was* a small but significant difference between black and white patients, that black people's airways responded uniquely differently to asthma.

The lead author of the paper herself tells me that her study can't explain higher asthma rates among black people overall, only the severity of their condition. But as Kaufman explains, their adjustment can't even necessarily explain this. "In a trial we would never adjust something that itself is affected by the exposure." It's like testing a drug with a known side effect of weight gain. You can't judge the effects of that drug by adjusting for weight, because the drug itself causes patients to gain weight. In the asthma study, "they're saying let's imagine a world in which black people don't have a more severe disease. Then what would their inflammatory response be? What use is that?" In Richard Cooper's words, the science is being retrofitted to accommodate race.

"People get trained in schools to build models and make adjustments. This is the way we do things. And then they just apply it to race as though it's the same as a pill you would take. It's completely bizarre," warns Kaufman. "Most practitioners of medical research with medical degrees and basic science degrees don't really have much background in statistics. Many people with perfectly good intentions end up committing a lot of statistical errors because of lack of training and something we call "wish bias," which is this idea that you want to find something interesting so you keep sifting through the data and fishing around until you find something interesting. That's a practice that generates many incorrect findings."

As useful as imaginary worlds can be, Kaufman tells me he cannot understand why medical researchers persist in applying statistical methods to race when it's obvious that the methods cannot work the way they want them to, that they produce imaginary worlds of little or no use. "It's epidemic in our literature that these adjustments are kind of nonsensical. In economics, this approach that epidemiologists and biomedical people use doesn't fly at all." He often sees scientists picking out a handful of variables to adjust for racial differences, without explaining why those ones were chosen and not others. At other times, he sees evidence of residual confounding, where the variables are measured poorly in the first place, making the final statistics even less reliable. This is especially true when it comes to adjusting for complex factors such as socioeconomic status.

The logic of statistical adjustment also holds true only if adjustment is actually possible. And racial difference is not a simple, measurable quantity like age or weight. The effects of racial discrimination, especially in a society as historically divided as the United States, run incalculably deep. It's not just about income or educational disparity. Black people in poorer neighborhoods live with worse levels of transportation, waste disposal, and policing, and they are more likely to have environmental hazards located nearby, such as bus garages, sewage treatment plants, and highways. These areas are also targets for cigarette and fast-food marketing.

Discrimination is a problem that extends all the way to the doctor's office. A hefty 432-page report published by the National Academy of Sciences in 2003 confirmed that evidence of racial and ethnic gaps in healthcare is "remarkably consistent across a range of illnesses and healthcare services," despite surveys showing that most Americans believe blacks and whites receive the same quality of care. More recently, Roosa Tikkanen, a researcher at the Commonwealth Fund, a private healthcare foundation based in New York, found that black and minority patients in New York and Boston are disproportionately treated at poorer, less well-equipped public hospitals rather than wealthier, high-quality teaching hospitals. She tells me that in some cases a public hospital may serve three times as many black patients as a private hospital located just a few blocks away.

Tikkanen suggests that institutional segregation and structural racism may be at play. Interviewing ambulance and emergency medical staff, she discovered that if someone was picked up from a low-income neighborhood, in the Bronx or a certain part of Brooklyn, and that person happened to be from a minority background, ambulance personnel would by default tend to take them to a public hospital or a handful of so-called "safety net" hospitals that have a history of welcoming poor and nonwhite patients. "Even after you account for the fact that they have worse insurance, minority patients are disproportionately seen in the public system," she adds.

In *The Social Life of DNA*, Columbia University sociologist Alondra Nelson explains that in eighteenth-century New York, the mortality rate of black infants was twice that of white. More than half the black population died in childhood, and those who lived had their bodies driven to the breaking point by the physical stresses of slavery. American medical experts claimed at the time that black people were naturally more robust, more resistant to diseases that killed others, including gallstones, tuberculosis, pneumonia, and syphilis. It was a narrative that served slavery, allowing slaveholders to subject people to harsher labor and living conditions on the assumption that this couldn't harm them. Nelson argues that this legacy of seeing people as intrinsically different, today manifested in lower living standards, has a direct impact on illness and death.

One 2018 study even found a possible relationship between racism in a geographical area and the health of newborns in that area. Researchers saw, astonishingly, a direct correlation between the proportion of Google searches for the N-word in an area and the prevalence of black babies being born prematurely or with a low birth weight. They also noted a similar heightened birth risk among women with Arab surnames in the six months after the 9/11 attacks. What is clear is that researchers are nowhere near understanding all the health impacts of the social factors around race.

In a paper published in the medical journal *Lancet* in 2017, a raft of public health researchers, including Mary Bassett, the New York City commissioner for health, warned that scientists were too often turning to biology to answer questions that could so clearly be better explained by social inequality. We know, for example, that 38 percent of non-Hispanic black children in the United States live below the poverty line, compared with less than 15 percent of children overall. Black Americans are clearly disadvantaged in employment and the legal system. In one study, in which an identical set of resumes was sent out to employers, white applicants with criminal records were called back more often than black applicants without criminal records.

"There is a rich social science literature conceptualizing structural racism, but this research has not been adequately integrated into medical and scientific literature," the authors of the study wrote. Structural racism is the elephant in the room. Of almost 48,000 articles they found on race and health, only 2,000 mentioned the word "racism" even once.

| | | | | | | | | |

The way data is collected and organized is also a problem.

Since 1993 the National Institutes of Health, the largest funder of medical research in the world, has had a general policy of requiring the clinical trials it supports to include women and minorities and also to collect data by race, across at least six categories. The purpose of this was never to look for differences between groups, but to ensure that medical studies include a broader spread of the population. As a demographic exercise, then, it made no difference that all black people, whether born in Africa or the United States, are today lumped by the National Institutes of Health into one general category, despite the prevalence of hypertension in their birth countries being completely different. It also shouldn't matter that the "white" category includes those from the Middle East and North Africa, as well as Northern Europe. Or that immigrants from Russia, which has generally very high levels of hypertension, are in the same category as white Americans, who tend to have far lower hypertension.

These were never considered to be genetically similar groups, just social and demographic ones. Collecting racial data like this was meant to be an exercise of ticking off boxes.

"All this is a record-keeping function that comports with other federal categories and guidelines, which are social categories," explains Dorothy Roberts, a law and sociology professor at the University of Pennsylvania. "They are the same categories as are used in the census to keep track of who is recruited into scientific studies. It's not a requirement for researchers to design their studies in any particular way." But that's not the way the data always ends up being used.

While it's important to know where and why racial disparities exist in health, the American habit of collecting data by race has the unintended consequence of driving researchers to use it, hunting for gaps and trying any means possible to explain those gaps. "The US government provides millions and millions of dollars of grant money targeted to this question. If the government is giving you money, saying we want you to answer this question, and this is all people know how to do, then they're going to answer that question whichever way they know how. So that provides some incentive," explains Jay Kaufman. It would be the same if funding agencies suddenly began collecting data by hair color as well as by race and sex. You would be almost certain to see studies suggesting biological differences between people with brown hair and blond hair. Just having the data invites comparisons.

But pooling people into groups is always imperfect, and the larger the group, the more imperfect it becomes. Environmental and cultural factors can play out in ways that outside observers may not anticipate.

For example, in Britain it's common for health researchers to lump all South Asian communities into one convenient category. The National Health Service website lists being South Asian as associated with a higher risk of cardiovascular disease. But this conclusion ignores the cultural and socioeconomic differences between Indians, Pakistanis, and Bangladeshis, even just within London. Rates of smoking tend to be high among Bangladeshis but very low among Indians, and smoking is a high-risk factor for cardiovascular disease. Vegetarianism is common among many Indians, but rare among Pakistanis, and diet is also a crucial component of cardiovascular disease. Hypertension is another condition also considered to be slightly higher among South Asians living in the UK. Yet despite the fact that India, Pakistan, and Bangladesh were one country until 1947, those of Indian origin tend to have higher blood pressure than those of Pakistani origin. Those of Bangladeshi ancestry have lower levels than Britain's white population.

With data as fuzzy and meaningless as this, adjusting for race or ethnicity can be a minefield, yet researchers routinely do it anyway. And sometimes with perplexing results, as Kaufman has found. One study he saw modeled hypertension and blood pressure for people in every country in the world in 1990 and in 2015. "So powerful was this model," he wrote, "that the authors even specified the mean blood pressure in 1990 for countries that did not even exist at that time."

| | | | | | | | |

The list of drugs available to treat hypertension reads like a Willy Wonka Chocolate Factory product catalogue: there are beta-blockers to slow down your heart rate, alpha-blockers to relax blood vessels, diuretics to shift salt out of your body, ACE-inhibitors that work on the hormones, calcium-channel blockers, vasodilators, and more. And within each of these categories are different brands, each competing fiercely for a piece of the lucrative blood pressure pie. They represent a global market worth about $80 billion. So it's small surprise that every pharmaceutical company wants to establish its unique selling point, to mark out its drug from the competition.

So it was with NitroMed, the marketers of BiDil, a pill that combined two different existing generic drugs, one that relaxed blood vessels and one that helped to combat heart failure, known to develop from hypertension. Around thirty years ago it became clear that mixing treatments in this way might make patients live longer, and BiDil was one of the earliest pills to do it. But there was a problem. The Food and Drug Administration refused to approve it because clinical trials hadn't been carried out fully, so the company was unable to show just how reliably it worked. Its patent was running out, leaving the company in a jam. How could they get their pill approved as quickly and cheaply as possible?

The solution they came up with was unprecedented: they carried out a clinical trial on black patients only.

Studies already suggested that black patients tended to respond a little less effectively to ACE-inhibitors, for reasons that weren't yet fully understood. It's a fact so widely accepted that hypertension pills in the United States often specify different usage guidelines for different races. In the

United Kingdom, clinical advice goes so far as to state that ACE-inhibitors should be given to white patients under the age of fifty-five to treat hypertension, but not to black patients. BiDil wasn't an ACE-inhibitor, which meant it might make a promising new first drug of choice for black patients, I'm told by Jay Cohn, the pill's developer, who is a cardiologist based at the University of Minnesota.

Cohn's original tests on BiDil had shown that the small number of black American patients who were included in the trial (just forty-nine people) seemed to respond better to the pill than other groups did. By 2004 the results of a new trial, this time carried out on around a thousand patients, were published in the *New England Journal of Medicine*; they confirmed that BiDil, when taken in addition to existing medication, reduced mortality rates by 43 percent. Given that the drug had already been shown to be effective in 1987, this outcome couldn't have come as a complete surprise. But what was certainly different this time was that every single one of the patients tested was black.

Cohn tells me that this was a practical decision. "We did not have adequate support to do a trial in the full population," he admits. The cost of full-scale clinical trials can easily run into many millions of dollars. "So we determined that maybe the best way to go would be to study the most responsive population, which was a self-designated black population." This didn't mean that BiDil didn't work in white patients, only that they didn't have the funds needed to do larger trials that included everyone. "We would have needed a larger sample size to study a general population than we could get away with, with a black population." On the basis of this one group-specific trial, in June the following year the Food and Drug Administration approved it as a medical treatment to be marketed solely to African Americans.

It was the world's first black pill.

As soon as the decision was made, it divided people. Health campaigners and some high-profile groups such as the Association of Black Cardiologists welcomed it as a positive move, signaling recognition, finally, of the historically neglected medical needs of black Americans. Others, including many doctors, saw it as little more than a cynical marketing ploy to leach more profit out of a drug on which the patent was about to expire. NitroMed, the company that owned the marketing rights to BiDil at the time, gained thirteen years of patent protection, and with this it could sell the combination drug for as much as six times the price it could charge for the individual pills separately. The pharmaceutical goldmine had been plumbed a little deeper.

That said, the going wasn't easy. NitroMed struggled to sell BiDil, in part because of skepticism among doctors, but also because of its eye-watering price. Since then, marketing rights have been sold to another pharmaceutical firm, Arbor Pharmaceuticals in Atlanta, Georgia. Go to its website and you'll see black models on nearly every page, smiling and reassuringly healthy-looking. "Although heart failure is on the rise all across America, it hits the African American community hardest," it states. There's no doubt that BiDil is still being marketed as a black pill.

Yet, as I'm reminded by Jonathan Kahn at the Mitchell Hamline School of Law in Minnesota, who has tracked the case from the beginning, "Race became relevant in the creation of this drug not for medical reasons, but for legal and commercial reasons." What's more, he explains, BiDil set a precedent. Seeing its success with the Food and Drug Administration, pharmaceutical firms began to file patent applications for other treatments that had been shown to work better in certain racial and ethnic groups. Looking at US patent applications filed between 2001 and 2005, the years running up to BiDil's approval, Kahn found that 65 mentioned race or ethnicity. Between 2006 and 2016, there were 384 that did.

Although no more explicitly race-specific drugs have been approved since BiDil, Kahn tells me that he has noticed an increased use of racial categories in drug labeling, the section that informs you about usage and dosage. "You know, Asians respond differently from Caucasians, that kind of language." For example, the website for the beta-blocker Bystolic highlights how well it works in black and Hispanic patients. In the first half of 2008, Bystolic was the most heavily advertised drug in the United States. Drug labeling might seem of little relevance, but, he explains, it is a "very powerful force for how medical professionals, and by extension anyone who's on these drugs, is being taught to think about the relationships of race to biology." When you see a drug specified for use in a certain group, it implies that this group of people is biologically different from others.

In reality those working in medical research know that race is hard to define; it is a poor proxy for how human variation really works. But when there are few easy ways to distinguish people, it can feel as good as any. The ultimate aim for many in medicine is not racialized medicine but personalized medicine, to be able to sequence an individual's genome and then tailor therapies to suit that individual. With personalized medicine, in principle nobody will ever need to take a drug that doesn't work on them, or that gives them a bad reaction. But sequencing everyone's individual genome is expensive and ethically fraught, and we don't yet have all the data we need to analyze them anyway. Given these limitations, grouping people by race is seen as an imperfect but practical approximation. Most doctors and medical researchers will admit that it's a fudge, but they use it anyway. A proxy can save money and time.

But how useful a fudge is race to medicine really? This is a question of statistics and demographics. Given all the medical data we have, how likely is it that an individual placed in a given racial group will benefit from a drug? And how likely is it that someone left out of that group could have benefited from it?

Take sickle cell anemia. It was once suggested that only black infants in the United States be screened for the sickle cell gene, because screening all infants would pose an unnecessary cost. Jay Kaufman and Richard Cooper worked together to dissect the statistics around sickle cell and found that, indeed, sickle cell trait prevalence in self-identified white Americans is only 250 per 100,000 members of the population, whereas in people who self-identify as black it is between 6,500 and 7,000. On this basis, it seems sensible to screen only black infants. On the other hand, there are many more white Americans than black Americans. The odds of a black newborn having the sickle trait may be 6.7 percent, but the odds of *any* newborn having it are in the same order of magnitude, around 1.5 percent. This is why US states today screen newborns universally, regardless of ethnicity or race.

For BiDil, this kind of number-crunching isn't possible because the 2004 clinical trials included only self-identified black Americans and no white Americans. Statistical comparison just couldn't be done. What we have instead is plenty of studies done over the decades on racial differences in response to common hypertension treatments. On the basis of these, the National Institute for Health and Care Excellence in the United Kingdom recommends treating black people with calcium-channel blockers as the drug of first choice, rather than ACE-inhibitors or other alternatives.

Kaufman and Cooper looked through all the published papers that studied responses to blood pressure medication. Their aim was to figure out just how many individuals actually benefit from this racial distinction. They discovered that the perceived racial differences in drug response are in fact relatively small compared with differences within racial groups—exactly to be expected, given everything scientists know about the genetics of human variation. So although there might be statistically significant differences at a population level, this isn't always useful when it comes to designing a treatment for any one individual patient. For example, they found that for ACE-inhibitors, which are given to white patients under the age of fifty-five in the United Kingdom but not to black patients, data suggest that for a hundred white people given the drug, forty-eight of them would fail to respond as hoped. Meanwhile if a hundred black people were given this drug they are usually denied, forty-one of them would benefit from it.

In this case, they conclude, assigning treatment by race is about as useful as flipping a coin.

| | | | | | | | | |

I ask Jay Cohn, the eminent cardiologist who invented BiDil, whether his drug, the world's first black pill, works well in patients who aren't black.

"Oh, of course it does! I use it all the time in white patients," he replies. "Everyone responds."

Cohn has known this all along, of course, and he has always been honest about it. His intentions were never to be racist, he tells me with a laugh. And I believe him. His goal was simply to get the drug approved any way he could. Labeling it as a "black pill" was only ever driven by a commercial imperative. And in the end, this is business. Indeed, pharmaceuticals are big business.

But patients who are prescribed different drugs or who read the labels telling them that race matters can find it difficult to parse just how complicated these facts really are. When it comes to big business, there are also those who may want to conceal facts. In January 2017 the independent news organization *ProPublica* revealed that Tom Price, a Republican congressman nominated by Donald Trump to become head of the Department of Health and Human Services, had persistently lobbied on behalf of Arbor Pharmaceuticals, which owns the marketing rights to BiDil, to remove a certain study from a government website. This study, carried out in 2009 by heart researchers at the University of Colorado on more than 76,000 people, showed that across all the racial and ethnic groups they studied, the combination of drugs that made up BiDil was not associated with significant reductions in mortality or hospitalization. The very first study on BiDil had looked at only 49 black Americans, and the second at just over a thousand. This study was far larger and therefore likely to be more significant.

The United States has a federal agency, the Agency for Healthcare Research and Quality, to help patients and doctors make informed choices about medical treatments. Yet according to the news report, one of Tom Price's aides had emailed the agency "at least half a dozen times" to have the University of Colorado study removed. It turned out that Arbor Pharmaceuticals had previously donated to Price's campaign fund. The month after the *ProPublica* news report came out, Price was confirmed as secretary of health and human services, although he resigned before the year was out, having been criticized for his use of expensive charter flights.

For now BiDil is still on the market, but the expiration date on its patent is coming up. In the meantime, race has become a firmly and widely accepted variable in medicine, according to the law professor Jonathan Kahn. He uses three clichés to describe what has happened in the last couple of decades. "The road to hell is paved with good intentions," is the first. "I don't think the use of race in these contexts was nefariously plotted," he says. Almost everyone in the medical research community believed they were doing the right thing, curing as many people as possible in the most efficient way.

The second is "the law of unintended consequences, resulting from that. And finally, that it's creating an accident waiting to happen." For Kahn, well-intentioned people have reintroduced race to medical science without fully understanding either their reasons or the consequences.

"It's not that the use of BiDil in black patients is immediately going to lead to the resubstantiation of slavery or scientific racism, but it's a step down the road of re-biologizing race in a way that feeds deep strains of racism," he tells me. "What we're seeing now in the US and again on the rise in Europe is a sort of ethno-nationalism. And any sort of indirect or direct approval or imprimatur of using race as a biological category becomes, no matter how well intended, dangerous in those contexts."

If the idea of race did no harm to anyone, perhaps there would be nothing wrong in making use of it sometimes, especially when it might at least provide some small kernel of usefulness. In the cash-strapped world of healthcare, why not?

But Evelynn Hammonds, the historian of science at Harvard University, warns me just how dangerous it is to wander down this road. The idea of race is not harmless. It brings with it centuries of political baggage, the blood of millions. And it has never been a neutral variable. "When Jesse Owens won the medals in the 1936 Olympics, some people argued that he was not a full-blooded Negro, that he was actually mixed, and so they measured his whole body. That was 1936. They measured his whole body and made arguments like this: the thighbone is within the normal range of what the normal Negroid thighbone should be," she says. "That's always the hot-button question at the end of the day. These people are different from each other, they're fundamentally different from each other, in disease, in athletic capacity, and, fundamentally in intelligence. That's the narrative. It acts as a kind of animating force under the surface."

Even when we know that there are population-level differences in disease frequency, blindly sticking to racial generalizations can be life-threatening. An American pediatrician, Richard Garcia, once described the case of a friend who repeatedly failed to receive a correct diagnosis for cystic fibrosis because it was thought to be a white disease, and she was black. Only when a passing radiologist happened to spot her chest x-ray, without knowing to whom it belonged, was her condition instantly spotted. She had to wait until she was eight years old, and her skin color had to be invisible, before she could be diagnosed.

| | | | | | | | | |

In 2003, Duana Fullwiley, an anthropologist then at the Harvard School of Public Health, spent six months watching medical researchers in laboratories in California. Their job was to find genetic differences in how people responded to drugs. It was a fairly young, diverse, international team, not at all stuffy or old-fashioned. And she noticed that all the scientists were routinely using racial categories not only to select their subjects but also to confidently pick out statistical differences between these racial groups.

So as Fullwiley observed them, she asked each scientist she interviewed one simple question: "How would you define race?"

Not one of them could answer her question confidently or clearly. The interviews were punctuated by long, awkward pauses and shy, embarrassed laughs. When pushed, some admitted that the concept of race made little sense, that the hard and fixed census categories actually didn't mean very much. One said, "You can only judge race to a certain degree of confidence." Another hesitated before admitting, "I need to think more about it."

Fullwiley concluded that most of the researchers "were unsure of the meanings of the race categories that they used, yet they continue to assert that there is a biological basis to them, which they will soon corroborate." Race was the entire premise upon which they were doing their research, but they were unable to tell her what it was. Their work instead seemed to rest upon a hope that if they just persisted, they would eventually come to find meaning in these categories. What they couldn't yet define would then be defined.

Somehow it would become real.

For all the studies that point to innate racial differences in health, the genetic evidence so far rarely tallies. Hypertension was just one case in point. Even enormous experiments looking at the genomes of thousands of people have turned up little. Although hundreds of gene variants linked to blood pressure have been found, collectively they explain just a percent or so of the variation we see, says Jay Kaufman. "We've had a decade of genome-wide association studies now, we've spent billions and billions of dollars, and we still are at the position that it looks like 97 percent of the mortality disparity between blacks and whites in the United States has nothing to do with genes."

And this makes sense, he adds. It would be bizarre to imagine that black Americans are somehow so uniquely biologically disadvantaged that they would naturally die of almost everything at higher rates than everyone else.

Dorothy Roberts agrees that there's no logic in expecting black Americans to be so medically unusual. "How could it possibly be that a group called black people, which first of all is defined differently around the world—it's been defined differently even within the United States, but the current definition in the United States is anyone with *any* discernible African ancestry—how could that hugely varied group, which could include someone with mostly European ancestry, someone with mostly Asian ancestry, someone with mostly Native American ancestry, how could it possibly be that that group could have a particular health outcome for an innate biological reason? That just doesn't make sense," she says and laughs. "The most plausible, to me the only possible, explanation could be because of inferior social conditions."

Kaufman suggests that part of the reason the United States clings to the idea of black exceptionalism when it comes to health may be because, in some way, it lets society off the hook. It lays the blame for inequality at the feet of biology. If poor health today is intrinsic to black bodies and has nothing to do with racism, it's not anyone's fault. "It says it's not about our organization of society that's somehow unfair or unjust or discriminatory. It's not that we treat people badly. It's not that we give people worse life chances. It's not that we are unfair or unjust as a society. It's just that these people have some genetic defect, and it's just the way they are."

During the era of slavery, recall, doctors proclaimed that blacks were uniquely tough, that they were resistant to pain and common diseases. "So you have at one time a literature saying that black people are especially robust, and at another time you have a literature saying that black people are especially predisposed to illness," he says. "It's a contradiction, but each one serves its own purpose."

Richard Cooper thinks that a psychological sea change is needed. What we think of as fact today will have to be replaced by a better understanding of race, just as what was accepted as fact yesterday has started to be replaced. Race is not a universal constant, a biological rule. "Race is a story we tell ourselves," he tells me. "Everybody has a general belief in race and then they have stories about it, either something they've seen or experienced. And the two reinforce each other. History, psychology, politics, we all have our belief systems and myths and things which a hundred years from now are not going to be true."

Yet we keep looking back to race because of its familiarity. We can't help it. For so long, it has been the backdrop to our lives, the running narrative. We automatically translate the information our eyes and ears receive into the language of race, forgetting where this language came from. "I think that scientists, they are trapped by the categories they use. They will either have to jettison it or find different ways of talking about this," says Hammonds. "They'll have to come to terms with that it has a social meaning." This doesn't mean that racial categories shouldn't be used in medicine or in science more generally. But it does mean that those who use them should fully understand what they mean, be able to define them, and to know their history.

They should at least know what race *is*.

# AFTERWORD

Barry Mehler's words ring in my ears.

"I have a lot of relatives who survived the Holocaust," the historian said. "They are prepared for things to cease to be normal very quickly because that was their experience."

I never thought I might live through times that could make me as anxious as this, that could also leave me dangling on a precipice afraid for my future. Politics is moving at such breakneck speed, taking such random turns, it seems anything is possible. It's the suddenness of it all that makes it feel so strange. The cancerous surge in nationalism and racism around the world has taken many of us by surprise. I grew up not very far from where a black teenager, Stephen Lawrence, was killed by racist thugs in Southeast London in 1993, while waiting for a bus. His murder left a mark on my generation. When we campaigned against racism, we knew there was a long way to go, but we were hopeful. And for a brief, sunlit moment things really did seem to be changing. My son was born five years ago, when Barack Obama was still the US president, and I dreamed that he might grow up in a better world, perhaps even a postracial one.

Things ceased to be normal very quickly.

In the space of just a few years, far-right and anti-immigrant groups have become visible and powerful across Europe and the United States. In Poland nationalists march under the slogan "Pure Poland, white Poland." In Italy a right-wing leader rises to popularity on the promise to deport illegal immigrants and turn his back on refugees. White nationalists look to Russia under Vladimir Putin as a defender of "traditional" values. In German elections in 2017 Alternative für Deutschland wins more than 12 percent of the vote. Steve Bannon, the former chief strategist to President Trump, tells far-right nationalists in France in 2018, "Let them call you racist, let them call you xenophobes, let them call you nativists. Wear it as a badge of honor."

While it may be easy to blame white supremacists for this cancer, it's a brand of identity politics that has others in its grip, too. It's infecting people everywhere, whether it's Islamic fundamentalists in the Middle East and Pakistan, Hindu nationalists in India, or Chinese scholars who turn their back on good science in favor of a worldview that paints the Chinese as having different evolutionary roots from everyone else. They may have different ideas and different histories, but their goal is the same: to assert difference for political gain. This is a twisted ideology that deliberately makes no appeal to a shared humanity, but instead rests on shadowy myths of belonging, on origin stories offering an umbrella to some but not others, sheltering them with false comfort. What nationalism stresses, as the late political scientist Ernst B. Haas wrote, is "the individual's search for identity with strangers in an impersonal world."

That desire to belong is powerful, I know. I was raised between cultures, and there's nothing quite as disruptive to your sense of belonging as not fully belonging anywhere, as being brown when everyone else is white in a place that notices these things. But don't be fooled. When people play on those feelings, when they tell you they can return you to a glorious past, offer a community of people just like you, who share your values and your dreams, a common history, they are selling you a myth. Enjoy your culture or religion, have pride in where you live or where your ancestors came from if you like, but don't imagine that these things give you any biological claim. Don't be sucked into believing that you are so different from others that your rights have more value, that your blood is a different color. There is no authenticity except the authenticity of personal experience.

The "race realists" as they call themselves (perhaps because calling yourself a racist is still unpalatable even to most racists) work so hard to make the opposite case. They appeal to that dark corner of our souls that wants to believe that human difference runs deep, making entire populations special, giving some nations an edge over others. And sadly, this is their moment. Whenever ugly politics become dominant, you can be sure that there are intellectuals and pseudointellectuals ready to jump on board. Those with dangerous ideas about "human nature" and even more dangerous prescriptions for our problems are always content to bide their time, knowing that the pendulum will swing their way eventually. Intellectual racism has always existed, and indeed for a chunk of history, it thrived. I believe it is still the toxic little seed at the heart of academia. However dead you might think it is, it needs only a little water, and now it's raining.

That said, what they're doing is also intellectually doomed. I've learned while writing this book that trying to force a biological understanding of race fails, often spectacularly, for the simple reason that history is the thing that can provide the answers. Science can't help you here. But then, perhaps the race realists know this. Maybe they know that if we truly want an end to racism, we need to understand the past, to have more equitable education and healthcare, to end discrimination in work and institutions, to be a little more open with our hearts and maybe also with our borders. Maybe they know that the answers are not in our blood, but they are in us. They are in our actions, in the choices we make, and in the ways we treat each other. Maybe their insistent banging of the drum, their increasing violence and anger, is simply to mask the fact that they don't want to make these concessions.

There are plenty of ignorant racists, but the problem is not just ignorance. The problem is that, even when people know the facts, not everyone actually wants an end to racial inequality. Some would rather things stayed the way they are, or even went backward. And this means that those committed to the biological reality of race won't back down if the data prove them wrong. There's no incentive for them to admit intellectual defeat. They will just keep reaching for fresher, more elaborate theories when the old ones fail. If skin color doesn't explain racial inequality, then maybe the structure of our brains and bodies will. If not anatomy, then maybe our genes. When then this, too, produces nothing of value, they will reach for the next thing. All this intellectual jumping through hoops to maintain the status quo. All this to prove what they have always really wanted to know: that they are superior.

Well, keep reaching, keep reaching. One day there will be nothing left to reach for.

# ACKNOWLEDGMENTS

This is the book I have wanted to write since I was a child, and I have poured my soul into it. I'm grateful that my editors at Fourth Estate and Beacon Press, Louise Haines and Amy Caldwell, didn't hesitate in commissioning it, and that my publicists Michelle Kane and Caitlin Meyer have been my champions throughout. My agents Peter Tallack, Louisa Pritchard, and Tisse Takagi were equally supportive. I am very fortunate to have such a loyal, kindhearted team around me.

I would also like to deeply thank Jon Marks, Eric Turkheimer, Bill Tucker, Jay Kaufman, Subhadra Das, Marek Kohn, Jennifer Raff, Greg Radick, and Billy Griffiths for their generous assistance. My friend the archaeologist Tim Power guided me through the British Museum and helped me see the past from a different perspective. I would also like to thank my sister Rima, herself a scholar of race and politics, for her critical feedback, and my husband, Mukul, for finding time to read chapters and for his patience while I was working. When we started dating, Mukul was a fan of the band Fun-Da-Mental, and he would often recite the two lines of lyrics I included at the beginning of this book.

In my business, there are true friends—the ones who care—and then there's everyone else. My deepest gratitude is reserved for Peter Wrobel, a true friend who out of the goodness of his heart scoured the entire manuscript for errors.

Writing is my second greatest pleasure. The first is my son, Aneurin. I don't know what the future holds, but I hope he never has to face the struggles that his parents did. I hope he understands that how we look, our genes, and even our distant ancestry are not the only things that give us our identities. Even culture is not everything. What makes us are our personal experiences and our individual actions.

Don't forget that, my little man.

# REFERENCES


## Prologue

Epigraph: From the album *Seize the Time*, by Fun-Da-Mental, released in 1994 on the Nation Records label. With deepest thanks to Dennis Webb for his kind permission to reproduce the words.

Factual details about the objects in the British Museum are taken from their labels, with guidance from archaeologist Tim Power.

"Sir Hans Sloane." British Museum website, www.britishmuseum.org/about_us/the_museums_story/general_history/sir_hans_sloane.aspx. Last accessed 19 November 2018.

John Bartlett. "'Stolen Friend': Rapa Nui Seek Return of Moai Statue." BBC News website, 18 November 2018. www.bbc.co.uk/news/world-latin-america-46222276.

Holger Hoock. "The British State and the Anglo-French Wars over Antiquities, 1798–1858." *Historical Journal* 50, no. 1 (March 2007): 49–72.

Johann Friedrich Blumenbach. *De generis humani varietate native*. Gottingae: Vandenhoek et Ruprecht, 1795.

Bernard J. Freedman. "Caucasian." *British Medical Journal* (Clinical Research Edition) 288, no. 6418 (3 March 1984): 696–98. https://uk.reuters.com/article/uk-africa-sarkozy/africans-still-seething-over-sarkozy-speech-idUKL0513034620070905.

MoveOn.org. "Cultural Genocide: U.S. Government Forces Egyptian Nubians to Be Classified as White and Not Black." Petition. https://petitions.moveon.org/sign/justice-for-an-indigenous. Last accessed 19 November 2018.

Shashi Tharoor. *Inglorious Empire: What the British Did to India*. London: Hurst & Company, 2017.

## Chapter 1: Deep Time

R. G. Gunn. "Mulka's Cave Aboriginal Rock Art Site: Its Context and Content." *Records of the Western Australian Museum* 23 (2006): 19–41.

Chris Clarkson et al. "Human Occupation of Northern Australia by 65,000 Years Ago." *Nature* 547 (20 July 2017): 306–10.

Sander van der Kaars et al. "Humans Rather Than Climate the Primary Cause of Pleistocene Megafaunal Extinction in Australia." *Nature Communications* 8, no. 14142 (20 January 2017).
Harry Allen. "The Past in the Present? Archaeological Narratives and Aboriginal History." In *Long History, Deep Time: Deepening Histories of Place*. Edited by Ann McGrath and Mary Anne Jebb, 176–202. Canberra, Australia: ANU Press, 2015.
Jean-Jacques Hublin et al. "New Fossils from Jebel Irhoud, Morocco and the Pan-African Origin of *Homo sapiens*." *Nature* 546 (8 June 2017): 289–92.
Steve Webb. *Made in Africa: Hominin Explorations and the Australian Skeletal Evidence*. Cambridge, MA: Academic Press, May 2018.
Kay Anderson and Colin Perrin. "'The Miserablest People in the World': Race, Humanism and the Australian Aborigine." *Australian Journal of Anthropology* 18, no. 1 (2007): 18–39.
Chris Gosden. "Race and Racism in Archaeology: Introduction." *World Archaeology* 38, no. 1 (2006): 1–7.
Billy Griffiths. *Deep Time Dreaming: Uncovering Ancient Australia*. Carleton, Victoria: Black Inc. Books, 2018.
Oodgeroo Noonuccal. "Stone Age." In *The Dawn Is at Hand*. London: Marion Boyars, 1992.
Bruce Pascoe. *Dark Emu, Black Seeds: Agriculture or Accident.* Broome, Western Australia: Magabala Books, 2014.
"Colonial Frontier Massacres in Central and Eastern Australia 1788–1930." University of Newcastle, Australia, website. https://c21ch.newcastle.edu.au/colonialmassacres/timeline.php. Last accessed 7 September 2018.
Commonwealth of Australia. *Bringing Them Home: Report of the National Inquiry into the Separation of Aboriginal and Torres Strait Islander Children from Their Families*. Sydney: Human Rights and Equal Opportunity Commission, 1997. www.humanrights.gov.au/sites/default/files/content/pdf/social_justice/bringing_them_home_report.pdf.
Meg Parsons. "Creating a Hygienic Dorm: The Refashioning of Aboriginal Women and Children and the Politics of Racial Classification in Queensland 1920s–40s." *Health and History* 14, no. 2 (2012): 112–39.
Martin Porr and Jacqueline M. Matthews. "Post-Colonialism, Human Origins and the Paradox of Modernity." *Antiquity* 91, no. 358 (August 2017): 1058–68.
Charles W. Mills. *The Racial Contract*. Ithaca, NY: Cornell University Press, 1997.
Tim Ingold. "Beyond Biology and Culture: The Meaning of Evolution in a Relational World." *Social Anthropology* 12, no. 2 (June 2004): 209–21.
Kay Anderson and Colin Perrin. "How Race Became Everything: Australia and Polygenism." *Ethnic and Racial Studies* 31, no. 5 (2008): 962–90.
Martin Porr. "Essential Questions: Modern Humans and the Capacity for Modernity." In *Southern Asia, Australia, and the Search for Human Origins*. Edited by Robin Dennell and Martin Porr, 257–64. Cambridge, UK: Cambridge University Press, 2014.
Johannes Krause et al. "The Complete Mitochondrial DNA Genome of an Unknown Hominin from Southern Siberia." *Nature* 464 (8 April 2010): 894–97.
Kay Prüfer et al. "The Complete Genome Sequence of a Neanderthal from the Altai Mountains." *Nature* 505 (2 January 2014): 43–49.
Ann Gibbons. "Who Were the Denisovans?" *Science* 333, no. 6046 (26 August 2011): 1084–87.

Alan G. Thorne and Milford H. Wolpoff. "Regional Continuity in Australasian Pleistocene Hominid Evolution." *American Journal of Physical Anthropology* 55, no. 3 (July 1981): 337–49.

Alan G. Thorne and Milford H. Wolpoff. "The Multiregional Evolution of Humans." *Scientific American* 266, no. 4 (April 1992): 76–83.

Gregory J. Adcock et al. "Mitochondrial DNA Sequences in Ancient Australians: Implications for Modern Human Origins." *Proceedings of the National Academy of Sciences* 98, no. 2 (January 2001): 537–42.

Jane Qiu. "The Forgotten Continent." *Nature* 535 (14 July 2016): 218–20.

Jin ChangZhu et al. "The Homo Sapiens Cave Hominin Site of Mulan Mountain, Jiangzhou District, Chongzuo, Guangxi, with Emphasis on Its Age." *Chinese Science Bulletin* 54, no. 21 (November 2009): 3848.

Fotios Alexandros Karakostis et al. "Evidence for Precision Grasping in Neandertal Daily Activities." *Science Advances* 4, no. 9 (26 September 2018).

Frances Wenban-Smith. "Neanderthals Were No Brutes—Research Reveals They May Have Been Precision Workers." *The Conversation*, 26 September 2018. https://theconversation.com/neanderthals-were-no-brutes-research-reveals-they-may-have-been-precision-workers-103858?utm_medium=Social&utm_source=Twitter#Echobox=1538043704. Last accessed 11 October 2018.

"Neanderthals Thought Like We Do." Max-Planck-Gesellschaft website. 22 February 2018. www.mpg.de/11948095/neandertals-cave-art. Last accessed 11 October 2018.

Emma Marris. "News: Neanderthal Artists Made Oldest-Known Cave Paintings." *Nature* 22 (February 2018).

Hélène Quach et al. "Genetic Adaptation and Neandertal Admixture Shaped the Immune System of Human Populations." *Cell* 167, no. 3 (20 October 2016): 643–56.

Michael Dannemann and Janet Kelso. "The Contribution of Neanderthals to Phenotypic Variation in Modern Humans." *American Journal of Human Genetics* 101, no. 4 (5 October 2017): 578–89.

Michael D. Gregory et al. "Neanderthal-Derived Genetic Variation Shapes Modern Human Cranium and Brain." *Scientific Reports* 7, no. 6308 (December 2017).

Annemieke Milks. "We Haven't Been Giving Neanderthals Enough Credit." *Popular Science*, 26 June 2018. www.popsci.com/neanderthal-hunting-spears.

"Neanderthals Were Too Smart for Their Own Good." *Telegraph*, 18 November 2011. www.telegraph.co.uk/news/science/science-news/8898321/Neanderthals-were-too-smart-for-their-own-good.html.

Sarah Kaplan. "Humans Didn't Outsmart the Neanderthals. We Just Outlasted Them." *Washington Post*, 1 November 2017. www.washingtonpost.com/news/speaking-of-science/wp/2017/11/01/humans-didnt-outsmart-the-neanderthals-we-just-outlasted-them/?utm_term=.4a78010999de.

Elizabeth Kolbert. "Our Neanderthals, Ourselves." *New Yorker*, 12 February 2015. www.newyorker.com/news/daily-comment/neanderthals.

Jon Mooallem. "Neanderthals Were People, Too." *New York Times*, 11 January 2017. www.nytimes.com/2017/01/11/magazine/neanderthals-were-people-too.html.

Eleanor M. L. Scerri et al. "Did Our Species Evolve in Subdivided Populations across Africa, and Why Does It Matter?" *Trends in Ecology & Evolution* 33, no. 8 (August 2018): 582–94.

Sheela Athreya. "Picking a Bone with Evolutionary Essentialism." *Anthropology News* (18 September 2018). www.anthropology-news.org/index.php/2018/09/18/picking-a-bone-with-evolutionary-essentialism.

**Chapter 2: It's a Small World**

Clifford Geertz. *The Interpretation of Cultures*. New York: Basic Books, 1973.

Charline Zeitoun. "In the Days of Human Zoos." CNRS website, 22 November 2016. https://news.cnrs.fr/articles/in-the-days-of-human-zoos.

Bernard Lewis. "The Historical Roots of Racism." *American Scholar* 67, no. 1 (Winter 1998): 17–25.

Kenan Malik. *Strange Fruit: Why Both Sides Are Wrong in the Race Debate*. London: Oneworld, 2008.

S. Solly, Geo Moojen, and Bernth Lindfors. "Courting the Hottentot Venus." *Africa: Rivista trimestrale di studi e documentazione dell'Istituto italiano per l'Africa e l'Oriente* 40, no. 1 (March 1985): 133–48.

"The Hottentot Venus Is Going Home." *Journal of Blacks in Higher Education*, no. 35 (Spring 2002): 63.

Jonathan Marks. *Is Science Racist?* Cambridge, UK: Polity Press, January 2017.

Daina Ramey Berry. *The Price for Their Pound of Flesh: The Value of the Enslaved, from Womb to Grave, in the Building of a Nation*. Boston: Beacon Press, 2017.

"The African-American Mosaic." Library of Congress website. www.loc.gov/exhibits/african/afam002.html. Last Accessed 12 October 2017.

Stephen Jay Gould. *The Mismeasure of Man*. New York: W. W. Norton, 1981.

Bronwen Douglas. "Climate to Crania: Science and the Racialization of Human Difference." In *Foreign Bodies: Oceania and the Science of Race 1750–1940*. Edited by Bronwen Douglas and Chris Ballard. Canberra, Australia: ANU Press, 2008.

Samuel A. Cartwright. "Diseases and Peculiarities of the Negro Race." *De Bow's Review*, Southern and Western States 11, New Orleans, 1851.

Arthur, Comte de Gobineau. *The Inequality of Human Races* (1853). Translated by Adrian Collins. London: William Heinemann, 1915.

John R. Swanton. "Review: The Inequality of Human Races by Arthur De Gobineau and Adrian Collins." *American Anthropologist*, New Series 18, no. 3 (July–September 1916): 429–31.

Conor Cruise O'Brien. "Thomas Jefferson: Radical and Racist." *Atlantic*, October 1996.

Adrian Desmond and James Moore. *Darwin's Sacred Cause: How a Hatred of Slavery Shaped Darwin's Views on Human Evolution*. New York: Houghton Mifflin Harcourt, 2009.

Gregory Radick. "Darwin and Humans." In *The Cambridge Encyclopedia of Darwin and Evolutionary Thought*. Edited by Michael Ruse, 173–81. Cambridge, UK: Cambridge University Press, 2013.

Gregory Radick. "How and Why Darwin Got Emotional About Race." In *Historicizing Humans: Deep Time, Evolution and Race in Nineteenth Century British Sciences*. Edited by Efram Sera-Shriar. Pittsburgh: University of Pittsburgh Press, 2018.

Thomas Henry Huxley. *Collected Essays: Volume 3, Science and Education*. London: Macmillan and Co., 1893.

Karen E. Fields and Barbara J. Fields. *Racecraft: The Soul of Inequality in American Life*, reprint edition. London: Verso, 2014.
Mitch Keller. "The Scandal at the Zoo." *New York Times*, 6 August 2006.

**Chapter 3: Scientific Priestcraft**

Susanne Heim, Carola Sachse, and Mark Walker. *The Kaiser Wilhelm Society Under National Socialism*. New York: Cambridge University Press, 2009.
James Hawes. *The Shortest History of Germany*. London: Old Street Publishing, 2017.
"History of the Kaiser Wilhelm Society Under National Socialism." Max-Planck-Gesellschaft website. www.mpg.de/9811513/kws-under-national-socialism. Last accessed 22 October 2017.
Katrin Weigmann. "In the Name of Science." *EMBO Reports* 2, no. 10 (2001): 871–75.
Benno Müller-Hill. "The Blood from Auschwitz and the Silence of the Scholars." *History and Philosophy of the Life Sciences* 21, no. 3 (1999): 331–65.
Kim Christian Priemel. "Review: The Kaiser Wilhelm Society under National Socialism, by Susanne Heim, Carola Sachse, and Mark Walker." *Journal of Modern History* 83, no. 1 (March 2011): 216–18.
Robert Wald Sussman. *The Myth of Race: The Troubling Persistence of an Unscientific Idea*. Cambridge, MA: Harvard University Press, 2014.
Marek Kohn. *The Race Gallery: The Return of Racial Science*. London: Jonathan Cape, 1995.
Annette Tuffs. "German Research Society Apologises to Victims of Nazis." *British Medical Journal* (16 June 2001): 322.
Michael Boulter. "The Rise of Eugenics, 1901–14." *Bloomsbury Scientists: Science and Art in the Wake of Darwin*. London: UCL Press, 2017.
Francis Galton. "Hereditary Character and Talent." *Macmillan's Magazine* 12 (1865): 157–66.
Siddhartha Mukherjee. *The Gene: An Intimate History*. New York: Scribner, 2016.
"The Eugenics Record Office." *Science* 37, no. 954 (11 April 1913): 553–54.
"International Eugenics Congress." *Scientific Monthly* 12, no. 4 (April 1921): 383–84.
"First International Eugenics Congress, London, July 24th to July 30th, 1912, University of London, South Kensington: Programme and Time Table," https://archive.org/stream/b22439833/b22439833_djvu.txt. Last accessed 12 January 2018.
Edwin Black. *IBM and the Holocaust: The Strategic Alliance Between Nazi Germany and America's Most Powerful Corporation*. New York: Crown, 2001.
Roswell H. Johnson. "Eugenics and So-Called Eugenics." *American Journal of Sociology* 20, no. 1 (July 1914): 98–103.
John R. Durant. "Scientific Naturalism and Social Reform in the Thought of Alfred Russel Wallace." *British Journal for the History of Science* 12, no. 1 (March 1979): 31–58.
Jason Burke and Philip Oltermann. "Germany Moves to Atone for 'Forgotten Genocide' in Namibia." *Guardian*. 25 December 2016.
Yuehtsen Juliette Chung. "Better Science and Better Race? Social Darwinism and Chinese Eugenics." *Isis* 105, no. 4 (December 2014): 793–802.
Prescott F. Hall. "Selection of Immigration." *Annals of the American Academy of Political and Social Science* 24 (July 1904): 169–84.

Jonathan H. X. Lee. *History of Asian Americans: Exploring Diverse Roots*. Santa Barbara, CA: Greenwood, 2015.
Madison Grant. *The Passing of the Great Race; Or the Racial Basis of European History*. New York: Charles Scribner's Sons, 1916.
A.B.S. "Reviewed Work: The Passing of the Great Race: Or the Racial Basis of European History by Madison Grant." *American Historical Review* 22, no. 4 (July 1917): 842–44.
Jonathan Spiro. *Defending the Master Race: Conservation, Eugenics, and the Legacy of Madison Grant*. Burlington: University of Vermont Press, 2009.
James Q. Whitman. *Hitler's American Model: The United States and the Making of Nazi Race Law*. Princeton, NJ: Princeton University Press, 2017.
Corey Johnson. "California Was Sterilising Its Female Prisoners as Late as 2010." *Guardian*, 8 November 2013.
Elliott Gabriel. "National Hygiene and 'Inferior Offspring': Japan's Eugenics Victims Demand Justice." *Mint Press News*, 19 March 2018. www.mintpressnews.com/inferior-offspring-japans-eugenics-victims-demand-justice/239078. Last accessed 26 March 2018.
Gavin Schaffer. "'Like a Baby with a Box of Matches': British Scientists and the Concept of 'Race' in the Inter-War Period." *British Journal for the History of Science* 38, no. 3 (September 2005): 307–24.

### Chapter 4: Inside the Fold

With thanks to King's College London for access to their special collections.

"Reginald Ruggles Gates Collection." King's College London website. www.kcl.ac.uk/library/archivespec/special-collections/Individualcollections/rugglesgates.aspx. Last accessed 2 November 2017.
Gavin Schaffer. "'Scientific' Racism Again? Reginald Gates, the 'Mankind Quarterly' and the Question of 'Race' in Science After the Second World War." *Journal of American Studies* 41, no. 2 (August 2007): 253–78.
Ashley Montagu. *Man's Most Dangerous Myth: The Fallacy of Race*. New York: Columbia University Press, 1942.
Ashley Montagu. "The Genetical Theory of Race, and Anthropological Method." *American Anthropologist* 44, no. 3 (July–September 1942).
UNESCO. "The Race Concept: Results of an Inquiry." *The Race Question in Modern Science*. Paris: UNESCO, 1952.
Richard Charles Lewontin. "The Apportionment of Human Diversity." *Evolutionary Biology* 6 (1972): 381–98.
A. W. F. Edwards. "Human Genetic Diversity: Lewontin's Fallacy." *Bioessays* 25, no. 8 (August 2003): 798–801.
D. J. Witherspoon et al. "Genetic Similarities Within and Between Human Populations." *Genetics* 176, no. 1 (May 2007): 351–59.
Noah A. Rosenberg et al. "Genetic Structure of Human Populations." *Science* 298, no. 5602 (20 December 2002): 2381–85.
Michael C. Campbell and Sarah A. Tishkoff. "African Genetic Diversity: Implications for Human Demographic History, Modern Human Origins, and Complex Disease Mapping." *Annual Review of Genomics and Human Genetics* 9 (2008): 403–33.

Perrin Selcer. "Beyond the Cephalic Index: Negotiating Politics to Produce UNESCO's Scientific Statements on Race." *Current Anthropology* 53, Supplement 5 (April 2012).

Veronika Lipphardt. "From 'Races' to 'Isolates' and 'Endogamous Communities': Human Genetics and the Notion of Human Diversity in the 1950s." In *Human Heredity in the Twentieth Century*. Edited by Bernd Gausemeier, Staffan Müller-Wille, and Edmund Ramsden, 55–68. London: Pickering & Chatto, 2013.

Rachel Silverman. "The Blood Group 'Fad' in Post-War Racial Anthropology." *Kroeber Anthropological Society Papers* 84 (2000): 11–27.

Indera P. Singh and Darshan Singh. "The Study of ABO Blood Groups of Sainis of Punjab." *American Journal of Physical Anthropology* 19, no. 3 (September 1961): 223–26.

Henry E. Garrett, "Klineberg's Chapter on Race and Psychology." *Mankind Quarterly* (July 1960): 15–22.

Henry E. Garrett. "The Equalitarian Dogma." *Mankind Quarterly* (April 1961): 253–57.

R. Gayre. "The Dilemma of Inter-Racial Relations." *Mankind Quarterly* 6, no. 4 (April–June 1966).

Božo Škerlj. "The Mankind Quarterly." *Man* 60 (November 1960): 172–73.

Santiago Genoves. "Racism and 'The Mankind Quarterly.'" *Science* 134, no. 3493 (8 December 1961): 1928–32.

William H. Tucker. *The Funding of Scientific Racism: Wickliffe Draper and the Pioneer Fund*. Urbana: University of Illinois Press, 2002.

Robert Wald Sussman. *The Myth of Race: The Troubling Persistence of an Unscientific Idea*. Cambridge, MA: Harvard University Press, 2014.

Geoffrey Ainsworth Harrison. "Reviewed Work: The Mankind Quarterly by R. Gayre of Gayre." *Man* 61 (September 1961): 163–64.

International Consortium of Investigative Journalists Offshore Leaks Database. https://offshoreleaks.icij.org/nodes/21000166. Last accessed 19 December 2017.

Tim Kelsey. "Ulster University Took Grant from Fund Backing Whites." *Independent*, 9 January 1994.

Adam Miller. "The Pioneer Fund: Bankrolling the Professors of Hate." *Journal of Blacks in Higher Education*, no. 6 (Winter 1994–95): 58–61 (a version of this article originally appeared in the *Los Angeles Times* in 1994).

Michael Kunzelman. "University Accepted $458K from Eugenics Fund." Associated Press, 25 August 2018. https://apnews.com/a9791e6174374437b3bbe17af8b76215.

Edward Dutton. "Obituary: Tatu Vanhanen 1929–2015." *Mankind Quarterly* 56, no. 2 (2015): 225–32.

David Barash. "Review: Race, Evolution and Behavior." *Animal Behaviour* 49, no. 4 (April 1995): 1131–33.

### Chapter 5: Race Realists

Barry Mehler. "The New Eugenics: Academic Racism in the U.S. Today." *Science for the People* 15, no. 3 (May–June 1983): 18–23.

Roger Pearson. "Immigration into Britain." *Northlander* 1, no. 1 (April 1958): 2.

Tim Kelsey and Trevor Rowe. "Academics 'Were Funded by Racist American Trust.'" *Independent on Sunday*, 4 March 1990.

Barry Mehler. "Rightist on the Rights Panel." *The Nation*, 7 May 1988, 640–41.

Frank Santiago. "Rights Official Has Racial 'Purity' Links." *Des Moines Register*, 28 February 1988.

Ralph Scott. "Arthur Jensen: A Latter-Day 'Enemy of the People'?" *Mankind Quarterly* 53, no. 3–4 (2013).

Richard Lynn. "The Intelligence of American Jews." *Personality and Individual Differences* 36, no. 1 (January 2004): 201–6.

Richard Lynn and Gerhard Meisenberg. "Review Article: The Average IQ of Sub-Saharan Africans: Comments on Wicherts, Dolan, and van der Maas." *Intelligence* 38, no. 1 (January–February 2010): 21–29.

Maciej Stolarski, Marcin Zajenkowski, and Gerhard Meisenberg. "National Intelligence and Personality: Their Relationships and Impact on National Economic Success." *Intelligence* 41, no. 2 (March–April 2013): 94–101.

Elsevier. "Intelligence—Editorial Board." Elsevier website. www.journals.elsevier.com/intelligence/editorial-board. Accessed 10 November 2017 and again 20 November 2018.

Arthur Jensen. "How Much Can We Boost IQ and Scholastic Achievement?" *Harvard Educational Review* 39, no. 1 (April 1969): 1–123.

Ben Van Der Merwe. "Exposed: London's Eugenics Conference and Its Neo-Nazi Links." *London Student*, 10 January 2018. http://londonstudent.coop/news/2018/01/10/exposed-london-eugenics-conferences-neo-nazi-links/amp/?__twitter_impression=true.

Charles A. Murray and Richard Herrnstein. *The Bell Curve: Intelligence and Class Structure in American Life*. New York: Free Press Paperbacks, 1996.

Charles Lane. "The Tainted Sources of 'The Bell Curve.'" *New York Review of Books*, 1 December 1994.

Steven J. Rosenthal. "The Pioneer Fund: Financier of Fascist Research." *American Behavioral Scientist* 39, no. 1 (September–October 1995): 44–61.

Eric Siegel. "The Real Problem with Charles Murray and 'The Bell Curve.'" *Scientific American* online. 12 April 2017. https://blogs.scientificamerican.com/voices/the-real-problem-with-charles-murray-and-the-bell-curve.

Robert Wald Sussman. "America's Virulent Racists: The Sick Ideas and Perverted 'Science' of the American Renaissance Foundation." *Salon*, 11 October 2014, www.salon.com/2014/10/11/americas_virulent_racists_the_sick_ideas_and_perverted_science_of_the_american_renaissance_foundation.

"A Convocation of Bigots: The 1998 American Renaissance Conference." *Journal of Blacks in Higher Education*, no. 21 (Autumn 1998): 120–24.

Louis Nelson. "Clinton Ad Ties Trump to KKK, White Supremacists." *Politico*, 25 August 2016. www.politico.com/story/2016/08/clinton-ad-kkk-trump-227404.

Nina Burleigh. "Steve Bannon, Jared Taylor and the Radical Right's Ivy League Pedigree." *Newsweek*, 23 March 2017. www.newsweek.com/bannon-spencer-trump-alt-right-breitbart-infowars-yale-gottfried-oathkeepers-572585.

Michael Levin. "Howard Beach Turns a Beam on Racial Tensions." *New York Times*, 11 January 1987, letter to the editor.

Southern Poverty Law Center. "Michael Levin." SPLC website. www.splcenter.org/fighting-hate/extremist-files/individual/michael-levin. Last accessed 13 November 2017.

Joel Stein. "Milo Yiannopoulos Is the Pretty, Monstrous Face of the Alt-Right." *Bloomberg*, 15 September 2016. www.bloomberg.com/features/2016-america-divided/milo-yiannopoulos.
Michael Schulson. "Race, Science, and Razib Khan." *Undark*, 28 February 2017. https://undark.org/article/race-science-razib-khan-racism.
Harry F. Weyher. "The Pioneer Fund, the Behavioral Sciences, and the Media's False Stories." *Intelligence* 26, no. 4 (1998): 319–36.

**Chapter 6: Human Biodiversity**

Jonathan Marks. *Human Biodiversity: Genes, Race, and History*. New York: Aldine de Gruyter, 1995.
Kenan Malik. *Strange Fruit: Why Both Sides Are Wrong in the Race Debate*. Oxford, UK: Oneworld, 2008.
Ron Unz. "How to Grab the Immigration Issue." *Wall Street Journal*, 24 May 1994. www.unz.com/runz/how-to-grab-the-immigration-issue.
Charles A. Murray and Richard Herrnstein. *The Bell Curve: Intelligence and Class Structure in American Life*. New York: Free Press Paperbacks, 1996.
Park MacDougald and Jason Willick. "The Man Who Invented Identity Politics for the New Right." *New York Intelligencer* (blog), 30 April 2017. http://nymag.com/daily/intelligencer/2017/04/steve-sailer-invented-identity-politics-for-the-alt-right.html.
Steve Olson. "The Genetic Archaeology of Race." *Atlantic*, April 2001. www.theatlantic.com/magazine/archive/2001/04/the-genetic-archaeology-of-race/302180.
Veronika Lipphardt. "'Geographical Distribution Patterns of Various Genes': Genetic Studies of Human Variation after 1945." *Studies in History and Philosophy of Biological and Biomedical Sciences* 47 (September 2014): 50–61.
WHO Scientific Group on Research in Population Genetics of Primitive Groups. *Research in Population Genetics of Primitive Groups: Report of a WHO Scientific Group (Meeting Held in Geneva from 27 November to 3 December 1962)*. Geneva: World Health Organization, 1964.
Luigi Luca Cavalli-Sforza. "The Human Genome Diversity Project: An Address Delivered to a Special Meeting of UNESCO." Paris, 12 September 1994.
Joanna Radin. "Human Genome Diversity Project: History." In *International Encyclopedia of the Social & Behavioral Sciences*, 2nd edition. Edited by James D. Wright, 306–10. Oxford, UK: Elsevier, 2015.
Matthew Walker. "First, Do Harm." *Nature* 482 (9 February 2012): 148–52.
Rob Evans. "Military Scientists Tested Mustard Gas on Indians." *Guardian*, 1 September 2007. www.theguardian.com/uk/2007/sep/01/india.military.
Gary Taubes. "Scientists Attacked for 'Patenting' Pacific Tribe." *Science* 270, no. 5239 (17 November 1995): 1112.
Jenny Reardon. *Race to the Finish: Identity and Governance in an Age of Genomics*. Princeton, NJ: Princeton University Press, 2004.
Luigi Luca Cavalli-Sforza. *Genes, Peoples, and Languages*. London: Penguin, 2001.
Lisa Gannett. "Racism and Human Genome Diversity Research: The Ethical Limits of 'Population Thinking.'" *Philosophy of Science* 68, no. 3 (2001): S479–S492.
UNESCO IBC Working Group on Population Genetics. *Draft Report: Bioethics and Human Population Genetics Research*. Third session of the IBC, 27–29 September 1995.

"Ancestry Names Margo Georgiadis Chief Executive Officer." Ancestry.com, 19 April 2018, www.ancestry.com/corporate/newsroom/press-releases/ancestry-names-margo-georgiadis-chief-executive-officer.

Henry Louis Gates. *Finding Oprah's Roots: Finding Your Own*. New York: Crown, 2007.

"Oprah Winfrey's Surprising DNA Test." Ancestry.com. https://blogs.ancestry.com/cm/the-surprising-facts-oprah-winfrey-learned-about-her-dna. Last accessed 7 March 2018.

Jenny Reardon. "Finding Oprah's Roots, Losing the World: Beyond the Liberal Anti-racist Genome." Unpublished paper presented at Berkeley Workshop on Environmental Politics, 23 October 2009.

Debbie A. Kennett et al. "The Rise and Fall of BritainsDNA: A Tale of Misleading Claims, Media Manipulation and Threats to Academic Freedom." *Genealogy* 2, no. 4 (2 November 2018).

Eric Boodman. "White Nationalists Are Flocking to Genetic Ancestry Tests. Some Don't Like What They Find." *STAT News*, 16 August 2017, www.statnews.com/2017/08/16/white-nationalists-genetic-ancestry-test.

**Chapter 7: Roots**

"Morrissey: Big Mouth Strikes Again." *New Musical Express*, December 2007.

Lucy Pasha-Robinson. "One in Three Black, Asian or Minority Ethnic People Racially Abused Since Brexit, Study Reveals." *Independent*, 17 March 2017. www.independent.co.uk/news/uk/home-news/one-three-black-asian-minority-ethnic-bame-racism-abuse-assault-brexit-hate-crime-tuc-study-a7634231.html.

University College London. "Face of First Brit Revealed." Press release. 7 February 2018. www.ucl.ac.uk/news/news-articles/0218/070218-Face-of-cheddar-man-revealed.

Selina Brace et al. "Population Replacement in Early Neolithic Britain." *bioRxiv*, 18 February 2018. www.biorxiv.org/content/early/2018/02/18/267443.

Brian Reade. "Dark-Skinned Cheddar Man Is Hard Cheese for the Racist Morons of the Far Right, Says Brian Reade." *Mirror*, 10 February 2018. www.mirror.co.uk/news/uk-news/dark-skinned-cheddar-man-hard-11999683.

Chauncey Devega. "Cheddar Man Is 'Black'! Another Racial Panic for White Supremacists." *Salon*, 12 February 2018. www.salon.com/amp/cheddar-man-is-black-another-racial-panic-for-white-supremacists.

Nicholas G. Crawford et al. "Loci Associated with Skin Pigmentation Identified in African Populations." *Science*, published online, 12 October 2017.

David Reich. *Who We Are and How We Got Here*. Oxford, UK: Oxford University Press, 2018.

Iain Mathieson et al. "Genome-Wide Patterns of Selection in 230 Ancient Eurasians." *Nature* 528, no. 7583 (December 2015): 499–503.

Iain Mathieson et al. "The Genomic History of Southeastern Europe." *Nature* online, 21 February 2018. www.nature.com/articles/nature25778.

Iñigo Olalde et al. "The Beaker Phenomenon and the Genomic Transformation of Northwest Europe." *Nature* 555 (8 March 2018): 190–96.

Morten E. Allentoft et al. "Population Genomics of Bronze Age Eurasia." *Nature* 522, no. 7555 (June 2015): 167–72.

Kristian Kristiansen et al. "Re-Theorising Mobility and the Formation of Culture and Language Among the Corded Ware Culture in Europe." *Antiquity* 91, no. 356 (April 2017): 334–47.

Tegwen Long et al. "Cannabis in Eurasia: Origin of Human Use and Bronze Age Trans-Continental Connections." *Vegetation History and Archaeobotany* 26, no. 2 (March 2017): 245–58.

Priya Moorjani et al. "Genetic Evidence for Recent Population Mixture in India." *American Journal of Human Genetics* 93, no. 3 (September 2013): 422–38.

Vagheesh M. Narasimhan et al. "The Genomic Formation of South and Central Asia." *bioRxiv*, 31 March 2018. www.biorxiv.org/content/early/2018/03/31/292581.

Adam Rutherford. *A Brief History of Everyone Who Ever Lived: The Stories in Our Genes.* London: Weidenfeld & Nicholson, 2016.

"Nobelist Jim Watson Is Honored on His 90th Birthday." Cold Spring Harbor Laboratory website, 11 April 2018. www.cshl.edu/nobelist-jim-watson-honored-90th-birthday.

James Watson. *The Double Helix: A Personal Account of the Discovery of the Structure of DNA*. New York: Atheneum, 1968.

Charlotte Hunt-Grubbe. "The Elementary DNA of Dr Watson." *Sunday Times*, 14 October 2007.

**Chapter 8: Origin Stories**

Qinglong Wu et al. "Outburst Flood at 1920 BCE Supports Historicity of China's Great Flood and the Xia Dynasty." *Science* 353, no. 6299 (August 2016): 579–82.

Ann Gibbons. "There's No Such Thing as a 'Pure' European—or Anyone Else." *Science* (15 May 2017).

Brendan D. O'Fallon and Lars Fehren-Schmitz. "Native Americans Experienced a Strong Population Bottleneck Coincident with European Contact." *Proceedings of the National Academy of Sciences* 108, no. 51 (December 2011): 20444–48.

Jennifer Raf. "Rejecting the Solutrean Hypothesis." *Guardian* online, 21 February 2018. www.theguardian.com/science/2018/feb/21/rejecting-the-solutrean-hypothesis-the-first-peoples-in-the-americas-were-not-from-europe.

Stephen Oppenheimer, Bruce Bradley, and Dennis Stanford. "Solutrean Hypothesis: Genetics, the Mammoth in the Room." *World Archaeology* 46, no. 5 (October 2014): 752–74.

John Lindo et al. "Ancient Individuals from the North American Northwest Coast Reveal 10,000 Years of Regional Genetic Continuity." *Proceedings of the National Academy of Sciences of the United States of America* 114, no. 16 (April 2017): 4093–98.

Maanasa Raghavan et al. "Upper Palaeolithic Siberian Genome Reveals Dual Ancestry of Native Americans." *Nature* 505 (January 2014): 87–91.

Michael J. O'Brien et al. "On Thin Ice: Problems with Stanford and Bradley's Proposed Solutrean Colonisation of North America." *Antiquity* 88, no. 340 (May 2014): 606–13.

Dennis Stanford and Bruce Bradley. "Reply to O'Brien et al." *Antiquity* 88, no. 340 (January 2015): 614–21.

Michael J. O'Brien et al. "Solutreanism." *Antiquity* 88, no. 340 (January 2015): 622–24.

Jennifer A. Raff and Deborah A. Bolnick. "Does Mitochondrial Haplogroup X Indicate Ancient Trans-Atlantic Migration to the Americas? A Critical Re-Evaluation." *PaleoAmerica* 1, no. 4 (November 2015): 297–304.

Jason Colavito. "White Nationalists and the Solutrean Hypothesis." *Blog*, 31 January 2014. www.jasoncolavito.com/blog/white-nationalists-and-the-solutrean-hypothesis.

"Nephi's Neighbors: Book of Mormon Peoples and Pre-Columbian Populations." Fair Mormon website. www.fairmormon.org/conference/august-2003/nephis-neighbors-book-of-mormon-peoples-and-pre-columbian-populations. Last accessed 23 March 2018.

Douglas Preston. "The Kennewick Man Finally Freed to Share His Secrets." *Smithsonian Magazine*, September 2014.

Kim TallBear. *Native American DNA: Tribal Belonging and the False Promise of Genetic Science*. Minneapolis: University of Minnesota Press, 2013.

Amy Harmon. "Indian Tribe Wins Fight to Limit Research of Its DNA." *New York Times*, 21 April 2010.

Jessica Bardill et al. "Advancing the Ethics of Paleogenomics." *Science* 360, no. 6387 (27 April 2018): 384–85.

Kim TallBear. "Genomic Articulations of Indigeneity." *Social Studies of Science* 43, no. 4 (August 2013): 509–33.

Morten Rasmussen et al. "The Genome of a Late Pleistocene Human from a Clovis Burial Site in Western Montana." *Nature* 506 (February 2014): 225–29.

Carl Zimmer. "New Study Links Kennewick Man to Native Americans." *New York Times*, 19 June 2015.

Tia Ghose. "Ancient Kennewick Man Finally Laid to Rest." *Live Science*, 22 February 2017. www.livescience.com/57977-kennewick-man-reburied.html.

Sara Jean Green. "'A Wrong Had Finally Been Righted': Tribes Bury Remains of Ancient Ancestor Known as Kennewick Man." *Seattle Times*, 19 February 2017.

Romila Thapar. *The Past as Present: Forging Contemporary Identities Through History*. New Delhi: Aleph Book Company, 2014.

Bettina Arnold. "The Past as Propaganda: Totalitarian Archaeology in Nazi Germany." *Antiquity* 64, no. 244 (September 1990): 464–78.

Bettina Arnold. "'Arierdämmerung': Race and Archaeology in Nazi Germany." *World Archaeology* 38, no. 1 (March 2006): 8–31.

Bernard Mees. "Hitler and Germanentum." *Journal of Contemporary History* 39, no. 2 (April 2004): 255–70.

David Crossland. "Germany Recalls Myth That Created the Nation." *Spiegel Online*, 28 August 2009. www.spiegel.de/international/germany/battle-of-the-teutoburg-forest-germany-recalls-myth-that-created-the-nation-a-644913.html.

"Use and Abuse of Ancient DNA." Editorial. *Nature* 555 (March 2018): 559.

Amy Harmon. "Geneticists See Work Distorted for Racist Ends." *New York Times*, 18 October 2018.

Rupam Jain and Tom Lasseter. "Special Report: By Rewriting History, Hindu Nationalists Aim to Assert Their Dominance over India." Reuters, 6 March 2018. www.reuters.com/investigates/special-report/india-modi-culture.

Romila Thapar. "Can Genetics Help Us Understand Indian Social History?" *Cold Spring Harbor Perspectives in Biology* 6, no. 11 (November 2014).

Akhilesh Pillalamarri. "When History Gets Political: India's Grand 'Aryan' Debate and the Indus Valley Civilization." *Diplomat*, 18 August 2016. https://thediplomat.com/2016/08/when-history-gets-political-indias-grand-aryan-debate-and-the-indus-valley-civilization.

Kai Friese. "4500-Year-Old DNA from Rakhigarhi Reveals Evidence That Will Unsettle Hindutva Nationalists." *India Today*, 31 August 2018. www.indiatoday.in/amp/magazine/cover-story/story/20180910-rakhigarhi-dna-study-findings-indus-valley-civilisation-1327247-2018-08-31.

Manimugdha S. Sharma. "Faking History Starts Online." *Times of India*, 19 November 2017. https://timesofindia.indiatimes.com/home/sunday-times/faking-history-starts-online/articleshow/61705453.cms.

Sushil Srivastava. "The Abuse of History: A Study of the White Papers on Ayodhya." *Social Scientist* 22, no. 5–6 (May–June 1994): 39–51.

D. N. Jha. "Against Communalising History." *Social Scientist* 26, no. 9–10 (September–October 1998): 52–62.

"Asifa Bano: The Child Rape and Murder That Has Kashmir on Edge," BBC News website, 12 April 2018. www.bbc.co.uk/news/world-asia-india-43722714.

**Chapter 9: Caste**

Jeiming Chen et al. "Genetic Structure of the Han Chinese Population Revealed by Genome-Wide SNP Variation." *American Journal of Human Genetics* 85, no. 6 (December 2009): 775–85.

Diane Coffey et al. "Explicit Prejudice." *Economic and Political Weekly* 53, no. 1 (6 January 2018).

"Father Kills Daughter for Marrying Outside Their Caste in Maharashtra." *Hindustan Times*, 6 April 2017. www.hindustantimes.com/india-news/father-allegedly-kills-daughter-for-marrying-outside-their-caste-in-maharashtra/story-WKLK054zk5IXznDJtThxnI.html.

Human Rights Watch. *"They Say We're Dirty"—Denying an Education to India's Marginalized.* 22 April 2014. www.hrw.org/report/2014/04/22/they-say-were-dirty/denying-education-indias-marginalized.

Melissa A. Ilardo et al. "Physiological and Genetic Adaptations to Diving in Sea Nomads." *Cell* 173, no. 3 (April 2018): 569–80.

Robert Plomin. *Blueprint: How DNA Makes Us Who We Are*. London: Allen Lane, 2018.

Lee Edson. "Jensenism: The Theory That I.Q. Is Largely Determined by the Genes." *New York Times Magazine*, 31 August 1969, 10.

William K. Stevens. "Doctor Foresees an I.Q. Caste System." *New York Times*, 29 August 1971.

J. Philippe Rushton and Arthur R. Jensen. "The Totality of Available Evidence Shows the Race IQ Gap Still Remains." *Psychological Science* 17, no. 10 (October 2006): 921–22.

Thomas J. Bouchard et al. "Sources of Human Psychological Differences: The Minnesota Study of Twins Reared Apart." *Science* 250, no. 4978 (12 October 1990): 223–28.

Constance Holden. "Identical Twins Reared Apart." *Science* 207, no. 4437 (21 March 1980): 1323–25 and 1327–28.

Daniel Jong Schwekendiek. "Height and Weight Differences Between North and South Korea." *Journal of Biosocial Science* 41, no. 1 (January 2009): 51–55.

Robert Plomin and Sophie von Stumm. "The New Genetics of Intelligence." *Nature Reviews Genetics* 19 (March 2018): 148–59.

Robert Plomin and Ian J. Deary. "Genetics and Intelligence Differences: Five Special Findings." *Molecular Psychiatry* 20 (2015): 98–108.

Suzanne Sniekers et al. "Genome-Wide Association Meta-Analysis of 78,308 Individuals Identifies New Loci and Genes Influencing Human Intelligence." *Nature Genetics* 49, no. 7 (July 2017): 1107–12.

H.-Hilger Ropers and Ben C. J. Hamel. "X-Linked Mental Retardation." *Nature Reviews Genetics* 6 (January 2005): 46–57.

Ariane Hegewisch and Emma Williams-Baron. "The Gender Wage Gap: 2017 Earnings Differences by Race and Ethnicity." Institute for Women's Policy Research, 7 March 2018. https://iwpr.org/publications/gender-wage-gap-2017-race-ethnicity.

Rakesh Kochhar and Anthony Cilluffo. "How Wealth Inequality Has Changed in the U.S. Since the Great Recession, by Race, Ethnicity and Income." Pew Research Center, 1 November 2017. www.pewresearch.org/fact-tank/2017/11/01/how-wealth-inequality-has-changed-in-the-u-s-since-the-great-recession-by-race-ethnicity-and-income.

Kids Count Data Center. "Children in Single-Parent Families by Race." https://datacenter.kidscount.org. Last accessed 4 December 2018.

Betty Hart and Todd R. Risley. *Meaningful Differences in the Everyday Experience of Young American Children*. Baltimore: Paul H. Brookes, July 1995.

"Against Bigotry." Editorial. *Nature* 548 (17 August 2017): 259.

James R. Flynn. "The Mean IQ of Americans: Massive Gains 1932 to 1978." *Psychological Bulletin* 95, no. 1 (1984): 29–51.

James R. Flynn. "Massive IQ Gains in 14 Nations: What IQ Tests Really Measure." *Psychological Bulletin* 101, no. 2 (1984): 171–91.

Richard E. Nisbett et al. "Intelligence: New Findings and Theoretical Developments." *American Psychologist* 67, no. 2 (February–March 2012): 130–59.

William T. Dickens and James R. Flynn. "Black Americans Reduce the Racial IQ Gap." *Psychological Science* 17, no. 10 (1 October 2006): 913–20.

Katarzyna Bryc et al. "The Genetic Ancestry of African Americans, Latinos, and European Americans across the United States." *American Journal of Human Genetics* 96, no. 1 (January 2015): 37–53.

Rebecca Gates-Coon. "The Children of Sally Hemings." Library of Congress website, May 2002. www.loc.gov/loc/lcib/0205/hemings.html.

P. A. Witty and M. A. Jenkins. "Intra-Race Testing and Negro Intelligence." *Journal of Psychology: Interdisciplinary and Applied* 1 (1936): 179–92.

Elsie G. Moore. "Family Socialization and the IQ Test Performance of Traditionally and Transracially Adopted Black Children." *Developmental Psychology* 22, no. 3 (May 1986): 317–26.

Andrew Colman. "Race Differences in IQ: Hans Eysenck's Contribution to the Debate in the Light of Subsequent Research." *Personality and Individual Differences* 103 (September 2016): 182–89.

Carl Cullinane and Philip Kirby. *Research Brief: Class Differences: Ethnicity and Disadvantage*. Sutton Trust, November 2016. www.suttontrust.com/wp-content/uploads/2016/11/Class-differences-report_References-available-online.pdf.

Nathaniel Comfort. "Book & Arts: Genetic Determinism Redux." *Nature* 561 (27 September 2018): 461–63.

Albert Einstein. *The Travel Diaries of Albert Einstein: The Far East, Palestine, and Spain, 1922–1923*. Edited by Ze'ev Rosenkranz. Princeton, NJ: Princeton University Press, 2018.

Marcus W. Feldman and Sohini Ramachandran. "Missing Compared to What? Revisiting Heritability, Genes and Culture." *Philosophical Transactions of the Royal Society B* 373, no. 1743 (5 April 2018).

Rama Shankar Singh. "The Indian Caste System, Human Diversity, and Genetic Determinism." In *Thinking About Evolution: Historical, Philosophical, and Political Perspectives, Vol. 2*. Edited by Rama Shankar Singh et al. Cambridge, UK: Cambridge University Press, 2000.

Rikhil R. Bhavnani and Alexander Lee. "Does Affirmative Action Worsen Bureaucratic Performance? Evidence from the Indian Administrative Service." Website of Rikhil R. Bhavnani, April 2018. https://faculty.polisci.wisc.edu/bhavnani/wp-content/uploads/2018/04/aa.pdf.

### Chapter 10: The Illusionists

Dean H. Hamer and Lev Sirota. "Beware the Chopsticks Gene." *Molecular Psychiatry* 5, no. 1 (February 2000): 11–13.

Michael Balter. "Brain Man Makes Waves with Claims of Recent Human Evolution." *Science*, New Series 314, no. 5807 (22 December 2006): 1871, 1873.

Patrick D. Evans et al. "Microcephalin, a Gene Regulating Brain Size, Continues to Evolve Adaptively in Humans." *Science*, New Series 309, no. 5741 (9 September 2005): 1717–20.

Nitzan Mekel-Bobrov et al. "Ongoing Adaptive Evolution of ASPM, a Brain Size Determinant in Homo sapiens." *Science*, New Series 309, no. 5741 (9 September 2005): 1720–22.

John Derbyshire. "Evolution of the Brain." *National Review*, 9 September 2005. www.nationalreview.com/corner/evolution-brain-john-derbyshire.

Gregory Cochran and Henry Harpending. *The 10,000 Year Explosion: How Civilization Accelerated Human Evolution*. New York: Basic Books, 2009.

Michael Balter. "Links Between Brain Genes, Evolution, and Cognition Challenged." *Science*, New Series 314, no. 5807 (22 December 2006): 1872.

J. Philippe Rushton et al. "No Evidence That Polymorphisms of Brain Regulator Genes Microcephalin and ASPM Are Associated with General Mental Ability, Head Circumference or Altruism." *Biology Letters* 3, no. 2 (2007): 157–60.

Bruce Lahn and Lanny Ebenstein. "Let's Celebrate Human Genetic Diversity." *Nature* 461, no. 461 (8 October 2009): 726–28.

Barbara Katz Rothman. *The Book of Life: A Personal and Ethical Guide to Race, Normality, and the Implications of the Human Genome Project*. Boston: Beacon Press, 2001.

Gregory Radick. "Presidential Address: Experimenting with the Scientific Past." *British Journal for the History of Science* 49, no. 2 (June 2016): 153–72.

Gregory Radick. "Beyond the 'Mendel-Fisher Controversy.'" *Science* 350, no. 6257 (9 October 2015): 159–60.

Richard C. Lewontin. "Biological Determinism." Tanner Lectures on Human Values 4, 1983, 147–83.

W. Carson Byrd and Victor E. Ray. "Ultimate Attribution in the Genetic Era: White Support for Genetic Explanations of Racial Difference and Policies." *Annals of the American Academy of Political and Social Sciences* 661, no. 1 (1 September 2015): 212–35.

Stephan Palmié. "Genomics, Divination, 'Racecraft.'" *American Ethnologist* 34, no. 2 (May 2007): 205–22.

James R. Flynn. *Asian Americans: Achievement Beyond IQ*. Hillsdale, NJ: Lawrence Erlbaum Associates, 1991.

British Medical Association. "Trend of Growing Numbers of BME Doctors in the Profession Continues." BMA website, 14 May 2018. www.bma.org.uk/about-us/equality-diversity-and-inclusion/equality-lens/trend-2.

Satoshi Kanazawa. "Why Are Black Women Less Physically Attractive Than Other Women?" *Psychology Today*, 15 May 2011. http://tishushu.tumblr.com/post/5548905092/here-is-the-psychology-today-article-by.

Satoshi Kanazawa and Raufhon Salahodjaev. "Why Do Societies with Higher Average Cognitive Ability Have Lower Income Inequality? The Role of Redistributive Policies." *Journal of Biosocial Science* 50, no. 3 (May 2018): 347–64.

Jelte M. Wicherts et al. "The Dangers of Unsystematic Selection Methods and the Representativeness of 46 Samples of African Test-Takers." *Intelligence* 38 (2010): 30–37.

Dylan Matthews. "Heritage Study Co-Author Opposed Letting In Immigrants with Low IQs." *Washington Post* online, 8 May 2013. www.washingtonpost.com/news/wonk/wp/2013/05/08/heritage-study-co-author-opposed-letting-in-immigrants-with-low-iqs/?utm_term=.cf882f04806f.

"US Migrant Children Cry for Separated Parents on Audio," *BBC News* website, 19 June 2018. www.bbc.co.uk/news/world-us-canada-44531187.

Eli Watkins and Abby Phillip. "Trump Decries Immigrants from 'Shithole Countries' Coming to US." *CNN Politics* website, 12 January 2018. https://edition.cnn.com/2018/01/11/politics/immigrants-shithole-countries-trump/index.html.

Rebecca Pinto et al. "Schizophrenia in Black Caribbeans Living in the UK: An Exploration of Underlying Causes of the High Incidence Rate." *British Journal of General Practice* 58, no. 551 (June 2008): 429–34.

Michael Balter. "Schizophrenia's Unyielding Mysteries." *Scientific American* (May 2017): 55–61.

Filippo Varese et al. "Childhood Adversities Increase the Risk of Psychosis: A Meta-Analysis of Patient-Control, Prospective- and Cross-Sectional Cohort Studies." *Schizophrenia Bulletin* 38, no. 4 (18 June 2012): 661–71.

Otmar von Verschuer. *Racial Biology of the Jews*. Translated by Charles E. Weber. Reedy, WV: Liberty Bell Publications, 1983.

### Chapter 11: Black Pills

Stephen J. Dubner. "Toward a Unified Theory of Black America." *New York Times Magazine*, 20 March 2005.

"Health A–Z: High Blood Pressure (Hypertension)." National Health Service website. www.nhs.uk/conditions/high-blood-pressure-hypertension. Last accessed 28 June 2018.

"Hypertension in Adults: Diagnosis and Management: Clinical Guideline." National Institute for Health and Care Excellence, published August 2011 and last updated November 2016. www.nice.org.uk/guidance/cg127/chapter/1-Guidance#initiating-and-monitoring-antihypertensive-drug-treatment-including-blood-pressure-targets-2. Last accessed 10 July 2018.

Keith C. Ferdinand et al. "Disparities in Hypertension and Cardiovascular Disease in Blacks: The Critical Role of Medication Adherence." *Journal of Clinical Hypertension* 19, no. 10 (October 2017): 1015–24.

Dorothy Roberts. *Fatal Invention: How Science, Politics, and Big Business Re-Create Race in the Twenty-First Century*. New York: New Press, 2012.

Richard S. Cooper. "Race in Biological and Biomedical Research." *Cold Spring Harbor Perspectives in Medicine* 3, no. 11 (November 2013).

"Global Health Observatory Data: Raised Blood Pressure." World Health Organization website. www.who.int/gho/ncd/risk_factors/blood_pressure_prevalence_text/en. Last accessed 10 July 2018.

"U.S. Public Health Service Syphilis Study at Tuskegee." Centers for Disease Control and Prevention website. www.cdc.gov/tuskegee/index.html. Last accessed 3 July 2018.

Richard S. Cooper et al. "An International Comparative Study of Blood Pressure in Populations of European vs. African Descent." *BioMed Central Medicine* 3, no. 2 (5 January 2005).

Katharina Wolf-Maier et al. "Hypertension Prevalence and Blood Pressure Levels in 6 European Countries, Canada, and the United States." *Journal of the American Medical Association* 289, no. 18 (14 May 2003): 2363–69.

Richard S. Cooper and Charles Rotimi. "Hypertension in Blacks." *American Journal of Hypertension* 10, no. 7 (1 July 1997): 804–12.

Jay S. Kaufman and Susan A. Hall. "The Slavery Hypertension Hypothesis: Dissemination and Appeal of a Modern Race Theory." *Epidemiology* 14, no. 1 (January 2003): 111–18.

Clarence E. Grim and Miguel Robinson. "Commentary: Salt, Slavery and Survival: Hypertension in the African Diaspora." *Epidemiology* 14, no. 1 (January 2003): 120–22.

Philip D. Curtin. "The Slavery Hypothesis for Hypertension Among African Americans: The Historical Evidence." *American Journal of Public Health* 82, no. 12 (December 1992): 1681–86.

Osagie K. Obasogie. "Oprah's Unhealthy Mistake." *Los Angeles Times*, 17 May 2007.

Adebowale Adeyemo et al. "A Genome-Wide Association Study of Hypertension and Blood Pressure in African Americans." *PLoS Genetics* 5, no. 7, published online 17 July 2009.

Srividya Kidambi et al. "Non-Replication Study of a Genome-Wide Association Study for Hypertension and Blood Pressure in African Americans." *BioMed Central Medical Genetics* 13, no. 27 (April 2012).

Amy L. Non et al. "Education, Genetic Ancestry, and Blood Pressure in African Americans and Whites." *American Journal of Public Health* 102, no. 8 (August 2012): 1559–65.

Pedro Ordúñez et al. "Blacks and Whites in the Cuba Have Equal Prevalence of Hypertension: Confirmation from a New Population Survey." *BioMed Central Public Health* 13, no. 169 (2013).
Joseph Tomson and Gregory Y. H. Lip. "Blood Pressure Demographics: Nature or Nurture . . . Genes or Environment?" *BioMed Central Medicine* 3, no. 3, 7 January 2005.
Richard S. Cooper et al. "Elevated Hypertension Risk for African-Origin Populations in Biracial Societies: Modeling the Epidemiologic Transition Study." *Journal of Hypertension* 33, no. 3 (March 2015): 473–81.
Feng J. He et al. "WASH—World Action on Salt and Health." *Kidney International* 78, no. 8 (2 October 2010): 745–53.
Jay S. Kaufman. "Statistics, Adjusted Statistics, and Maladjusted Statistics." *American Journal of Law & Medicine* 43 (May 2017): 193–208.
Elizabeth Arias et al. "United States Life Tables." National Vital Statistics Reports of the United States Centers for Disease Control and Prevention 66, no. 4 (August 2017).
Sharmilee M. Nyenhuis et al. "Race Is Associated with Differences in Airway Inflammation in Patients with Asthma." *Journal of Allergy and Clinical Immunology* 140, no. 1 (January 2017): 257–65.
"Asthma and African Americans." US Department of Health and Human Services Office of Minority Health website. https://minorityhealth.hhs.gov/omh/browse.aspx?lvl=4&lvlid=15. Last accessed 10 July 2018.
Jay S. Kaufman and Richard S. Cooper. "Use of Racial and Ethnic Identity in Medical Evaluations and Treatments." In *What's the Use of Race? Modern Governance and the Biology of Difference*. Edited by Ian Whitmarsh and David S. Jones. Cambridge, MA: MIT Press, 2010.
Jay S. Kaufman et al. "Socioeconomic Status and Health in Blacks and Whites: The Problem of Residual Confounding and the Resiliency of Race." *Epidemiology* 8, no. 6 (November 1997): 621–28.
Brian D. Smedley et al. *Unequal Treatment: Confronting Racial and Ethnic Disparities in Health Care*. Institute of Medicine Committee on Understanding and Eliminating Racial and Ethnic Disparities in Health Care. Washington, DC: National Academies Press, 2003.
Alondra Nelson. *The Social Life of DNA: Race, Reparations and Reconciliation After the Genome*. Boston: Beacon Press, 2016.
David H. Chae et al. "Area Racism and Birth Outcomes Among Blacks in the United States." *Social Science & Medicine* 199 (February 2018): 49–55.
Zinzi D. Bailey et al. "Structural Racism and Health Inequities in the USA: Evidence and Interventions." *Lancet* 389, no. 10077 (8 April 2017): 1453–63.
"NIH Policy and Guidelines on the Inclusion of Women and Minorities as Subjects in Clinical Research," US National Institutes of Health website. https://grants.nih.gov/grants/funding/women_min/guidelines.htm. Last accessed 20 August 2018.
National Health Service. "Cardiovascular Disease." NHS website. www.nhs.uk/conditions/cardiovascular-disease. Last accessed 9 July 2018.
Jawad M. Khan and Gareth D. Beevers. "Management of Hypertension in Ethnic Minorities." *Heart* 91, no. 8 (2005): 1105–9.
Transparency Market Research. "Global Cardiovascular Drugs Market: Incessantly Rising Cases of Hypertension and Hyperlipidemia to Fuel Market Growth." PR

Newswire, 7 March 2018. www.prnewswire.com/news-releases/global-cardiovascular-drugs-market-incessantly-rising-cases-of-hypertension-and-hyperlipidemia-to-fuel-market-growth-says-tmr-676097783.html. Last accessed 17 July 2018.

Anne L. Taylor et al. "Combination of Isosorbide Dinitrate and Hydralazine in Blacks with Heart Failure." *New England Journal of Medicine* 351, no. 20 (11 November 2004): 2049–57.

Jonathan D. Kahn. "Beyond BiDil: The Expanding Embrace of Race in Biomedical Research and Product Development." *St. Louis University Journal of Health Law & Policy* 3 (2009): 61–92.

Jonathan D. Kahn. "Race in a Bottle." *Scientific American* 297, no. 2 (August 2007): 40–45.

Leszek Kalinowski et al. "Race-Specific Differences In Endothelial Function: Predisposition Of African Americans to Vascular Diseases." *Circulation* 109, no. 21 (1 June 2004): 2511–17.

Pamela Sankar and Jonathan Kahn. "BiDil: Race Medicine or Race Marketing?" *Health Affairs* (July–December 2005). Web exclusive: W5-455-63.

Jonathan D. Kahn. "Revisiting Racial Patents in an Era of Precision Medicine." *Case Western Reserve Law Review* 67, no. 4 (2017): 1153–69.

Karl E. Hammermeister et al. "Effectiveness of Hydralazine/Isosorbide Dinitrate in Racial/Ethnic Subgroups with Heart Failure." *Clinical Therapeutics* 31, no. 3 (March 2009): 632–43.

Robert Faturechi. "When a Study Cast Doubt on a Heart Pill, the Drug Company Turned to Tom Price." *ProPublica*, 19 January 2017. www.propublica.org/article/when-a-study-cast-doubt-on-heart-pill-the-drug-company-turned-to-tom-price.

Richard S. Garcia. "The Misuse of Race in Medical Diagnosis." *Pediatrics* 111, no. 5 (May 2004): 1394–95.

Duana Fullwiley. "Race and Genetics: Attempts to Define the Relationship." *BioSocieties* 2, no. 2 (2007): 221–37.

### Afterword

Ian Bremmer. "These 5 Countries Show How the European Far-Right Is Growing in Power." *Time*, 13 September 2018. http://time.com/5395444/europe-far-right-italy-salvini-sweden-france-germany.

"Poland Independence: Huge Crowds March amid Far-Right Row," *BBC News* website, 11 November 2018. www.bbc.co.uk/news/world-europe-46172662.

Alexander Stille. "How Matteo Salvini Pulled Italy to the Far Right." *Guardian*, 9 August 2018. www.theguardian.com/news/2018/aug/09/how-matteo-salvini-pulled-italy-to-the-far-right.

Tom Porter. "Charlottesville's Alt-Right Leaders Have a Passion for Vladimir Putin." *Newsweek*, 16 August 2017. www.newsweek.com/leaders-charlottesvilles-alt-right-protest-all-have-ties-russian-fascist-651384.

Allison Kaplan Sommer. "The Global Anti-Globalist: Steve Bannon Comes Out as Proud 'Racist' on His European Comeback Tour." *Haaretz*, 11 March 2018. www.haaretz.com/us-news/.premium-on-european-tour-steve-bannon-comes-out-as-a-proud-racist-1.5890885.

Ernst B. Haas. "Review: What Is Nationalism and Why Should We Study It?" *International Organization* 40, no. 3 (Summer 1986): 707–44.

# INDEX

## ABOUT THE AUTHOR

Angela Saini is an award-winning science journalist and broadcaster based in London. She presents science programs for the BBC, and her writing has appeared in leading publications worldwide, including the *Guardian*, the *Times*, *Science*, and *Wired*. Her last book, *Inferior: How Science Got Women Wrong and the New Research That's Rewriting the Story*, was published in 2017 to widespread critical acclaim and was named the *Physics World* Book of the Year. Saini has a master's degree in engineering from the University of Oxford and is a former fellow of the Massachusetts Institute of Technology. Her journalism has received prizes from both the American Association for the Advancement of Science and the Association of British Science Writers.